Cartographic Communication

SCIENCES

Geography and Demography, Field Director – Denise Pumain

Cartography, Subject Head – Colette Cauvin-Reymond

Cartographic Communication

Graphic Semiology, Semiotics and Geovisualization

Coordinated by

Boris Mericskay

iSTE

WILEY

First published 2023 in Great Britain and the United States by ISTE Ltd and John Wiley & Sons, Inc.

ISTE Ltd
27-37 St George's Road
London SW19 4EU
UK

www.iste.co.uk

John Wiley & Sons, Inc.
111 River Street
Hoboken, NJ 07030
USA

www.wiley.com

Library of Congress Control Number: 2023930947

British Library Cataloguing-in-Publication Data
A CIP record for this book is available from the British Library
ISBN 978-1-78945-091-0

ERC code:
SH2 Institutions, Values, Environment and Space
SH2_11 Human, economic and social geography
SH2_12 GIS, spatial analysis; big data in political, geographical and legal studies

Contents

Foreword

Colette CAUVIN-REYMOND
LIVE, CNRS, University of Strasbourg, France

Within the framework of the ISTE SCIENCES encyclopedia and, more specifically, of the "geography and demography" field, four works are devoted to cartography. Cartography is a scientific and artistic discipline (Robinson 1952, 1953), and it is indispensable to any person, organization or institution that needs to process and represent geographic data to bring out its spatial characteristics. With current technological changes, the continuous increase in the use of Web 2.0 and the appearance of social networks, cartography is undergoing profound changes.

This discipline has, in fact, undergone many more or less fundamental changes over time, both conceptually and technically, the two being intimately linked, as de Rosnay (2008) writes: "Scientific progress and technological progress feed each other." If its junction with statistics, as early as the end of the 19th century, had introduced new processes to translate localized phenomenon, the advent of the computer, and subsequently the birth of computerized cartography, could be considered a revolution. The latter was coined in an article by Tobler in 1959, simultaneously announcing the steps for the construction of an automated map and the basic principles of what would become geographic information systems (GIS).

Between this date and the beginning of the 1980s, this "new" cartography was developed in two directions: on the one hand, computer cartography, which reproduced what had previously been done manually, and, on the other hand, computer-assisted cartography, which opened up new avenues for creating innovative representations or introducing methods of surface analysis[1]. In the early 1980s, microcomputers appeared, whose immense capacities are now available to a large number of people with minimum knowledge. "Making" maps seems to be easier, updates become simpler, software, of varying quality, is multiplying. The availability of the Internet, and especially the web in 1994, further simplified the diffusion of graphic documents, which multiplied.

In addition to these new features, a fundamental change involved the communication paradigm, which is essential in the field. Despite automation, the arrival of the Internet and the web, the logic of moving from an author/producer (the one who conceives, who represents) to a reader/user, sometimes with feedback, remains predominant; the reader, the one who consults, who looks at the map and uses it, remains passive, even if, thanks to animation and especially interactivity, he/she can move his document around, zoom over it, fly over it, etc.; but he cannot modify it. However, with the arrival of Web 2.0, the semantic web, and the advent of the Geoweb from 2005 to 2010, and especially 2015, a new turning point is taking place: the reader/user, whoever they may be, becomes active. They can modify the maps on the web and can even create them. This is the beginning of a new period for cartography, which will be further developed in the works for the ISTE encyclopedia, in order to present an updated state of this science, with its transformations for the period 2010–2021.

The chosen date of 2010 as the threshold is explained by the fact that many works on cartography, both in French (Béguin and Pumain 1994; Cauvin et al. 2010, 2008, 2010) and in English (Slocum et al. 2009), were published in the previous years, exposing the characteristics of the field just before the "real" start of the semantic web, highlighting the steps necessary to produce a map, as well as the successive choices that must necessarily be made. Indeed, the construction of a map requires at least three main steps, each of which has a very specific role. The first step is to create a localized database from the geographic information provided. The second stage ensures the processing and transformation of the data in the database, insisting either on the locations or on the thematic data, or on both simultaneously, with or without taking time into account; it leads to the determination of the representation mode adopted. As for the third stage, that of communication and diffusion, it is not only based on technical solutions, on semiotic choices, on the adoption of particular actions, etc., but also on the knowledge of elements of visual perception and cognition.

1 This approach largely corresponds to the analytical cartography introduced by Tobler (1976, 2000).

These three stages[2] are interrelated and are linked to both the challenges and the relationship between the recipient/user of the map. These three stages guided the selection of the four works chosen for the ISTE encyclopedia, because they start from the state of cartographic science at a period just before important changes. It will therefore be easier to highlight, describe and characterize the specific and original features of the years 2010–2021. One volume is devoted to each of the stages, and a fourth shows cartography from a historical aspect, thus facilitating a better understanding of the changes over time. Each volume can be read independently of the others and in an order that the reader wishes, according to his or her desires and expectations. The same is true for each chapter.

The aim of the historical volume is "to outline the history of cartography as it is done, and as it continues to evolve, by proposing a synopsis of the reflections and their modifications over the last forty years, in order to advance a history of cartography that takes into account current reflections and research, and above all, that opens up new avenues to explore."

The volume on geographic information and cartography focuses on data, its characteristics and its use in cartography during a period where digital techniques are gradually, but fundamentally, recomposing contemporary societies. Data acquisition is no longer an activity reserved for specialists. The sources of geodata are diversifying, citizens are directly and often voluntarily providing geographic information (GPS readings, cell phones, connected mobile objects, Twitter, etc.); networking is immediate. The search for information for the production of maps is therefore profoundly transformed and reveals major issues in terms of digital society.

The volume on the processing and mapping of geographic information presents techniques and methods for processing both spatial and thematic information, in order to produce maps adapted to each problem. Although the book first mentions traditional procedures that are still in use, it focuses on spatial and spatiotemporal processes, as well as their apprehension through indicators and models, specifying the original cartographic representations that result from them.

The volume on communication, however, deals with the transmission of the map, because the latter is not a document that we produce so that it can remain in our pocket. It must become visible, and eventually audible and sensible, by appealing to graphic variables or others types of variables, possible actions and presenting display device properties. This volume will certainly be the easiest for

2 A similar pattern is found in the latest version of Robinson's reference book from 1995.

detecting the specificities of the current period, with the role of choices at all stages, the interweaving of these stages and the collaborative work.

With the same tools and the same means, specialists and non-specialists will produce maps by taking what is at their disposal but with distinct training, which induces risks. Indeed, the new cartographic processes that are constantly emerging have many positive aspects: ease of use, diversity, attractiveness. However, like all new developments, their advantages have a downside; each of these processes has specific properties that are important to know, at least in order to apply them wisely and avoid mistakes. The four books presented here show that we are moving toward new logics, new ways of conceiving, translating, processing and cartographically transcribing geographic information, without necessarily having to reject what already exists. The world of maps does not escape the general evolution in which we are all participating, with our wealth and our worries. May these books help everyone to be aware of what they produce and to use all these tools with discernment.

References

Béguin, M. and Pumain, D. (1994). *La représentation des données géographiques. Statistique et cartographie*. Cursus, Paris.

Cauvin, C., Escobar, F., Serradj, A. (2010). *Thematic Cartography 1: Thematic Cartography and Transformations*. ISTE Ltd, London, and John Wiley & Sons, New York.

Cauvin, C., Escobar, F., Serradj, A. (2010). *Thematic Cartography 2: Cartography and the Impact of the Quantitative Revolution*. ISTE Ltd, London, and John Wiley & Sons, New York.

Cauvin, C., Escobar, F., Serradj, A. (2010). *Thematic Cartography 3: New Approaches in Thematic Cartography*. ISTE Ltd, London, and John Wiley & Sons, New York.

Robinson, A.H. (1952). The look of maps. An examination of cartographic design. PhD Thesis, University of Wisconsin, Madison.

Robinson, A.H. (1953). *Elements of Cartography*. Wiley, New York.

Robinson, A.H., Morrison, J., Muehrcke, P.C., Kimerling, A.J., Gutpill, S.C. (1995). *Elements of Cartography*. Wiley, New York.

Rosnay, J. (2008). *2020 : les scenarios du futur. Comprendre le monde qui vient*. Fayard, Paris.

Slocum, T.A., McMaster, R.B., Kessler, F.C., Howard, H.H. (2009). *Thematic Cartography and Geography Visualization*. Prentice Hall, Hoboken.

Tobler, W.R. (1959). Automation and cartography. *The Geographical Review*, 49(4), 526–534.

Tobler, W.R. (1976). Analytical cartography. *The American Cartographer*, 3, 21–31.

Tobler, W.R. (2000). The development of analytical cartography. *Cartography and Geographic Information Science*, 27(3), 189–194.

Introduction

Boris MERICSKAY
ESO, University of Rennes 2, France

From the first Ptolemy maps to Google Maps, through Cassini, Mercator and Brunet maps, cartography has evolved and changed with the times, uses and technological advances. However, regardless of its medium, its form, its designer or its context of use, a map always has the same objective: to simplify reality in order to transmit information and communicate a message.

With the development of computer science, the Internet and the web, cartography has profoundly evolved as a science, an art and a discipline. At the heart of the daily practices of both individuals and organizations, maps are becoming dynamic, interactive and participatory, producers and users of maps are diversifying, and cartographic instrumentation is becoming more and more accessible to a wider audience. Their production and dissemination methods are changing, and the traditional distinction between the producer and the receiver is becoming blurred, to the extent that the stages in the design of a map are becoming confused (data, processing, formatting, dissemination, etc.).

This new sociotechnical context is not without consequences with regard to the cartographic communication process, which needs to be questioned and considered in light of the new dynamics at play. At a time when communication through maps is becoming more widespread in multiple forms, when geographic data are circulating and when the professions that use spatial representations are multiplying and diversifying, it seems necessary to establish the basis for contemporary and transversal reflection on this question. This is the perspective adopted for this volume, which proposes keys to reading and understanding the issues related to cartographic communication, semiology and geovisualization in the digital and Geoweb era.

Structured in seven chapters, the book offers an overview of the cartographic communication issue by mobilizing theoretical, conceptual and methodological contributions from various fields of research (information and communication sciences, cartography, geomatics, computer science, data visualization, geovisualization, etc.). Through a series of additional approaches (semiological, semiotic, technical, esthetic, cognitive, etc.), it offers an overview of the uses and communication issues that maps have, at a time when cartography is becoming more popular.

Throughout the different chapters, the map is not considered as a simple representation of space, but as the result of a process of construction and mediation that mobilizes a whole set of concepts, methods and techniques. The questions around sign systems (semiotics), as well as graphic systems (semiology), are widely mobilized and examined. As for the methodological and technical aspects of cartographic communication, they are approached from various operational points of view (education, research, publishing, decision support, consultation) and considered through the spectrum of the different channels of production and dissemination of today's maps (CAD, GIS, web mapping, geovisualization, statistical software, etc.) The different authors in this book are part of different currents of thought, research themes and technical and methodological practices around the representation of spatial data.

The first four chapters propose a reading of cartographic communication with regard to technological developments, uses and audiences, focusing on questions of graphic semiology, modes of representation and models of communication.

To open the volume, Chapter 1 (Christine Zanin) questions the articulations between rules and attraction in contemporary cartographic communication. It provides an overview of the foundations and issues involved in cartographic communication by reviewing different facets of the cartographer that coexist and evolve in a rapidly changing technical context.

Continuing this general reflection, Chapter 2 (Laurent Jégou) examines the traditional model of cartographic communication (sender-receiver) by modifying and developing it according to a heuristic and multidisciplinary approach. Through an iterative work of deconstruction and progressive enrichment of the model, this contribution proposes a renewed reading of cartographic communication, in light of the sociotechnical evolutions of the last few years.

More focused on the practice of thematic cartography, Chapter 3 (Nicolas Lambert, Timothée Giraud and Ronan Ysebaert) deepens cartographic communication through an exercise of cartographic multi-representation. Using the same dataset of the world population, the chapter presents and discusses a series of 13 different maps in order to illustrate how cartography is, first and foremost, a question of choice and methods.

In this line of thought, Chapter 4 (Françoise Bahoken) specifically addresses mapping flows and movements proposing a rich summary of this issue, which requires specific reflections. This contribution returns to the elements of construction of these particular types of maps, as well as the associated communication issues in light of technical developments. The chapter also presents a multi-representation cartographic work on container exchange flows, which allows for a summary of the elements discussed.

The last three chapters broaden the spectrum of reflection by focusing on cartographic communication within new digital mapping environments (Geoweb, geovisualization).

Chapter 5 (Matthieu Noucher) presents cartographic communication through new ways of making and using maps in the Geoweb era. This thought process does not put the emphasis on maps, but on spatial data, whose production and use modalities are transformed within these online environments. Notions of volunteered geographic information, producer-consumer or ego-cartography are discussed, transforming the traditional models of cartographic communication.

Chapter 6 (Boris Mericskay) also deals with the Geoweb but from the perspective of thematic cartography, focusing on the new properties of web maps and the "new" cartographic language that is taking shape on the web. This chapter proposes a summary of the principles, tools and modes of representation that are taking place on the Geoweb and broadly deals with the question of technique and tools in the cartographic communication process.

Lastly, Chapter 7 (Sidonie Christophe) completes this work by addressing the articulations between geovisualization and cartographic communication that are taking place today. This original contribution addresses the issues of style, personalization and interaction at the center of new cartographic communication issues.

This book provides an overview of the different approaches used to improve the understanding and transmission of spatial information through maps. It highlights the fundamental role of technological changes, which impose a permanent reflection on anyone using cartographic communication for the production of static maps, web maps and geovisualization applications.

1

The Multiple Facets of the Cartographer: Communication between Rules and Attraction

Christine ZANIN
Géographie-cités, University of Paris-Diderot, France

1.1. Introduction

The issue of cartographic representation, as both a geographer's tool and as an image of a territory, is a rather trivial issue that tends to appear outdated. And yet, despite the habit of "seeing" maps, the place and role of cartography in society are becoming more complex and thus need to be reexamined. Ptolemy, the oldest of the geographer-cartographers, considered a map to be an object for showing "things we cannot see". In the modern era, the cartographer's competence to achieve an accurate measurement of the terrain is recognized as a fundamental characteristic of scientific cartography which, in the constant improvement of this measurement, presents a path toward the "perfect" and objective representation of the world (Guilhot 2005). Today, we recognize that a map cannot be objective, that there is no single graphic answer to a geographic problem and that it is, above all, an object of communication of territorial issues.

The issues surrounding cartographic communication are complex and numerous, and the cartographer tends to get lost between what is pure communication, which is the prerogative of cartography, and all the considerations related to the plethora of tools and methods. It is no longer just a matter of seeking a more or less objective (re)transcription of reality, but more of an aid to spatial reflection, to the

Cartographic Communication,
coordinated by Boris MERICSKAY. © ISTE Ltd 2023.

construction of a project, to the achievement of a spatialized objective. Besse (2010) assures us that "any map, directly or indirectly, is a matter and expression of project", whereas Gould (1995) and Harley (1989) have always considered it as an instrument of construction or deconstruction of power. For Harley, maps impose the dominant values and beliefs of a society, what Foucault calls "an act of surveillance". They believe that the abstract nature of maps tend to "de-socialize" the territory they represent, resulting in a rejection of their social aspect in favor of their scientific aspect.

The multiplication and technicalization of methods and tools for representing geographic data transform the map into a protean object, and make the cartographer's task ever more complex and ambiguous. Although we continue to try and locate ourselves, to situate ourselves in space, it is not necessarily useful to resort to a so-called "scientific" cartography to be effective in the transmission of messages. Without being a specialist, or trained in its principles, anyone can easily find their position in space, as well as produce and share their own maps. Smartphones, GPS, social networks: there are many new tools, especially mobile ones, that contribute to making the places where we live, our territories, our travels and our itineraries permanently public (Tobelem Zanin 2013).

The cartographic issue is therefore no longer an issue about clearing some *terra incognita*, but rather about the cartographic capacity to rethink the world around us. The power inherent in cartographic knowledge makes maps, to use Jacob's (1996) wording, "a semiological trap and an ideological weapon", with an astonishing capacity to give an objective and natural appearance to what is a cultural and social construction. Jacob (1992) proposes a definition of cartography in which "the map is essentially considered as a symbolic mediation between man and their spatial environment, but also between individuals who can communicate thanks to this visual support" (quoted by Guilhot 2005). Regardless of the definition and the approach chosen, cartography is no longer a simple "graphic language" but a medium, a communication tool (Figure 1.1), where multiple theoretical and scientific issues are at stake (Tobelem Zanin 2013).

Between complexity and simplicity, between estheticism and semiological rigor, between rules and attraction, the objective of communication must remain at the center of the cartographer's concerns. To this end, this chapter proposes looking further into the reflection on cartographic construction and to question it through different facets of the profession. We try to follow in the footsteps of the cartographer in order to understand how they think, construct and give meaning to their representations.

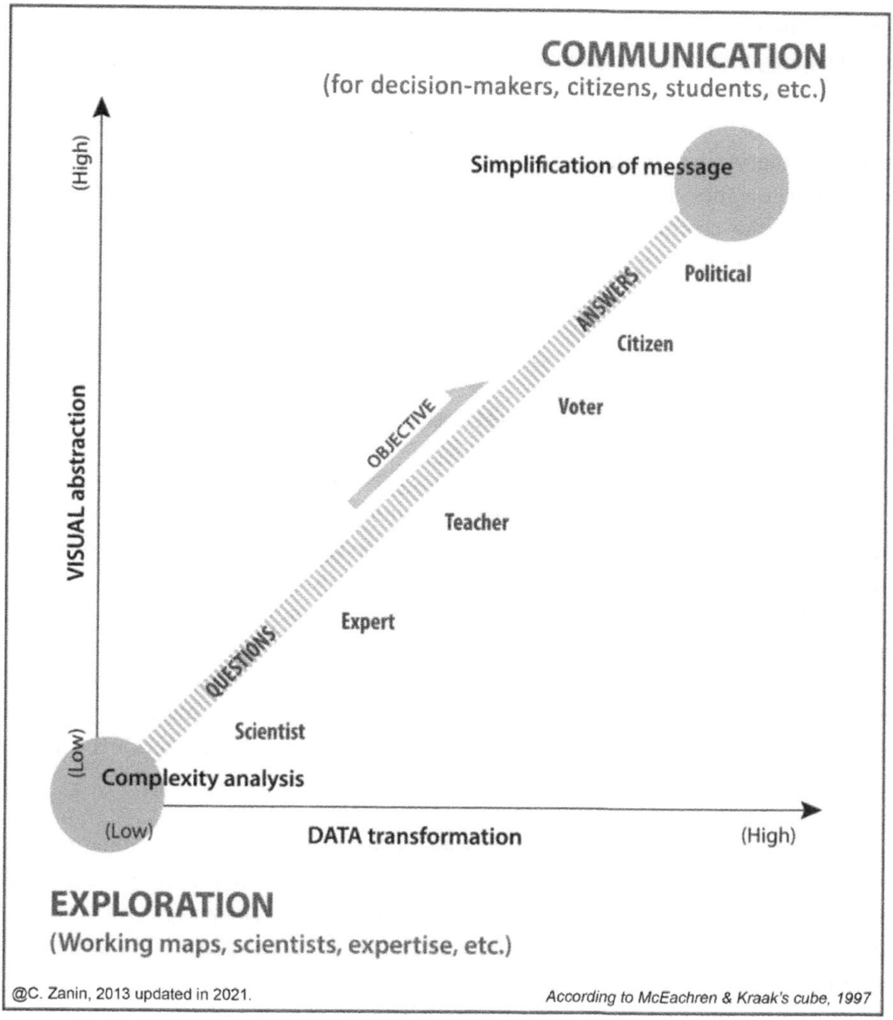

Figure 1.1. *The objective at the center of the representation*

The cartographer's first objective is to "show a graphical and spatial transcription of geographical information" (Tobelem Zanin 2013). However, the cartographer being a character with multiple skills and facets, the map is neither unique nor universal. The cartographer is a geographer, but not just that. They are the result of a clever mix between a geographer, a thematician, a mathematician, a computer scientist, an analyst, a technician, a manager, a planner and a mediator, without forgetting a certain artistic touch. The image remains at the center of their cartographic concern, but a rigorous construction of the image is not what gives all its meaning to their representation. We must add the power of the art of communication: the quality of what surrounds the image is what allows it to achieve visual strength in the message conveyed. This chapter thus takes the form of a review of the various facets of the cartographer, which we consider to be assets to communicate a message. Without any hierarchical order, from the semiologist to the artist, from the craftsman to the technician, from the thematic border-crosser to the communicator, from the computer scientist to the drawer, we will see how the cartographer declines their skills to become a multi-faceted character in the art of spatial representation.

1.2. The cartographer semiologist: how to use the map

The first facet of the cartographer that we consider is that of the semiologist. Unlike a writer, who conveys their message with words, the cartographer expresses themself through the specific cartographic language, of which three elementary graphic elements (the point, the line and the area) make up the alphabet and the visual variables the vocabulary (Figure 1.2). Of course, they make good use of grammar, allowing them to organize the chosen visual variables, but they must also master its syntax, which is based on the set of rules of visual perception. The introduction of the notion of visual perception radically changes the way the map is conceived. The meaning given to the map is perceived by a direct link between the signs perceived on the map and their translation into an idea.

More simply, we could follow the trail of the semiologist who explains that a sign has a signifying side (the form of the sign) and a signified side (what the sign expresses). The whole cartographic debate lies in the adequacy or dissonance between these two sides. In France, Bertin (1967) theorized this debate by establishing graphic semiology[1]. In the United States, Robinson (1952) worked at the same time on map design. As Palsky (2011) points out, there is "no concrete evidence of reciprocal influence, nor any correspondence between Arthur Robinson and Jacques Bertin", nevertheless both seek to "facilitate the passage of information and [...] reduce errors of misinterpretation. [...] They both concentrated on the elements of the map, the signs, which they considered as composing a system, a real language [...]".

1 For more information on graphic semiology's place in the cartographic communication process, see Chapter 2.

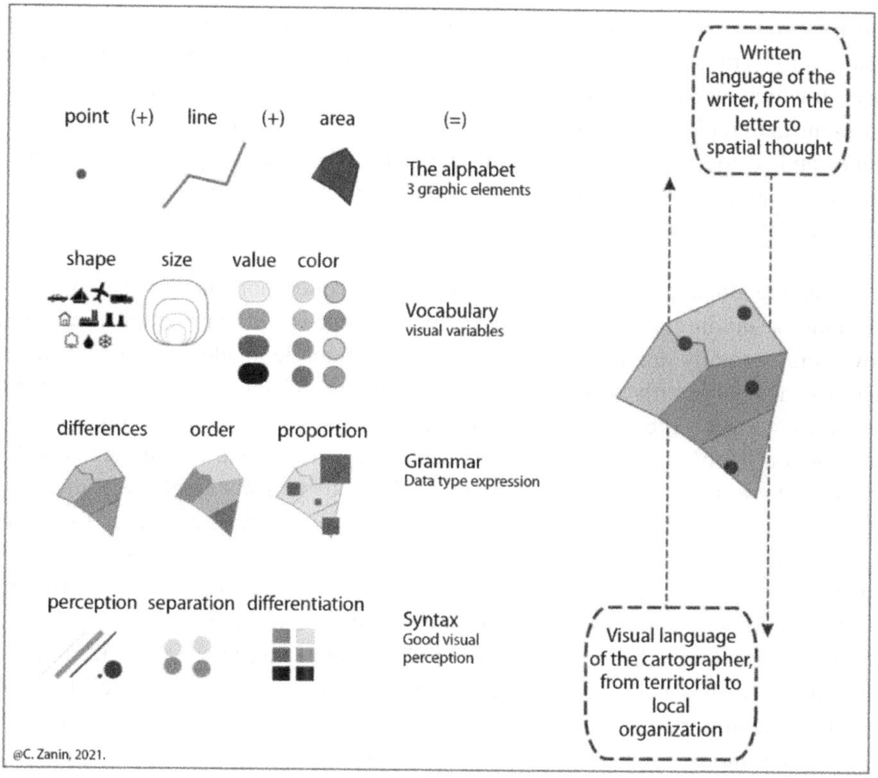

Figure 1.2. *The cartographic language. For a color version of this figure, see www.iste.co.uk/mericskay/cartographic.zip*

This research is completed by the works of MacEachren (1994, 1995) and Slocum (2005), in search of new semiological expressions induced by the introduction of dynamics[2] in spatial representations (MacEachren 1994; Andrienko 1997; Kraak 1998; Krygier 2011), through perception analyses (Jacob 2003; Fabrikant 2015; Maggi 2015; Christophe 2016; Ory 2016). The technological (r)evolutions that have followed one another since the appearance of microcomputing have totally changed the practice, as well as the reading of maps, whatever their nature[3]. "New approaches", "unavoidable" and "renewed

2 Research on so-called dynamic visual variables should not be confused with research on the dynamics of change, in other words, problems related to the representation of movement and change in time and space (Bahoken 2016).

3 Chapters 5 (on the Geoweb), 6 (on online cartography) and 7 (on geo-visualization) explore this issue extensively from different perspectives.

transformations" and "new paths to explore"[4] are on the agenda for contemporary cartography. Transformations in map design and display are definitely changing the ways in which maps are approached, and are greatly increasing the possible ways to represent geographic information. A reader who looks at a map tries to recognize shapes and memorize the most striking elements. "This capacity generates a focus, thus a selective attention and a concentration of consciousness, that is to say, a mental effort" (Cauvin et al. 2008).

Changing display and representation methods does not change that which is essential: a guiding thread of thought in the process of cartographic creation. However, it totally transforms the mental effort to be made. The eye and the memory act differently toward the understanding of the forms and territorial dynamics to be revealed. The semiological approach is based on rules of construction of the symbolism and is based on a codified use of the elements, but also on artistic principles. Each element of the map will always have a meaning, either because it is supporting the message, or by its mere presence, which forces the reader to interpret it, even if no particular meaning has been attributed to it. To transform a fact into a sign is not an insignificant and neutral operation. As Barthes (1985) points out, semiotics is also the science of levels: the informative level, the communicative level, the symbolic level and the signifying level. We will translate this into the different levels of reading the map, or the degree to which the user perceives the importance of the message conveyed by the map and transmitted by an adapted semiology; Bertin's (1977) "what stands out" implies that the graphical signs on the map make immediate sense.

Lastly, it is one of the main challenges of cartographic work to strive for balance in the use of graphic signs to arbitrate between the clarity of the signs used and the relevance of the links established as well as the points of view adopted. Everything is a matter of dialogue between objectives that are often contradictory. If the map is a source of information and a source of sensation, it must above all make sense, providing a basis for understanding. There are no good or bad maps, just more or less effective maps, which may or may not achieve their assigned objective when the balance is found.

The cartographer is not just a semiologist. The construction of the graphic sign to which they are committed also makes them a craftsman, or even an artist.

4 Titles of books by Cauvin et al. (2008), Volumes 1, 2, 4 and 5 (these volumes are also available in English as *Thematic Cartography 1-3*, published by ISTE Ltd and Wiley in 2010).

1.3. The cartographer, a craftsman drawer

The second facet of the cartographer is that of the craftsman, not quite an artist, not quite a scholar. The cartographer seeks to rid themself of a label that clings to them, that of the "nice digital technology specialist" (Tobelem Zanin 2013). They want to prove that if cartography is a practice and an art, it also has its place as a science. This battle is not new. We can already find traces of it in the writings of Robinson (1952), Rimbert (1968), as well as in the latest writings of Bertin (2004), who explains that the 20th-century cartographer is often mistaken for a printer, and then for a computer scientist.

Artist, craftsman or scientist, the training of the "map maker" (the title given to the 18th-century cartographer) has long been a serious matter, a journeyman matter. In the 20th century, its history merges with the direction of the Institute of Geography by de Martonne. In their 1934 annual report, they justified the creation of the School of Cartography by the need to "train cartographers-geographers at a university that would be able to apply their expertise" to all sectors where maps were needed (government services, press, tourism). The training of high-level cartographers was thus undertaken by the university. It combined "technical skill and geographical education", by marrying "the school of drawing and university education".

> Drawing was taught at length by Professor L. Bergelin and was based on topographic, geological, orohydrographic and geomorphological maps, wall maps and summary maps. One learns to draw the letter, to point a "picot" to a twentieth of a millimeter [...]. (Robic 1997; quoted by Bianchin 2012, p. 14)

We are here in an artistic and technical profession. The journeymen cartographers acquire their letters of nobility and are, from then on, part of a school of arts and crafts, a recognized and specific school. Technical skill and geographical culture: the cartographer is not just a "tinkerer", but a person who is able to think about geography, to make it express structures and to show territorial organizations in images (Figure 1.3). For Bertin (2004), the cartographer is one who is "capable of constructing original maps and of guiding the work of drawers who only have manual skills". The cartographer "must play a leading role in all public and private service mapping workshops" (de Martonne 1934).

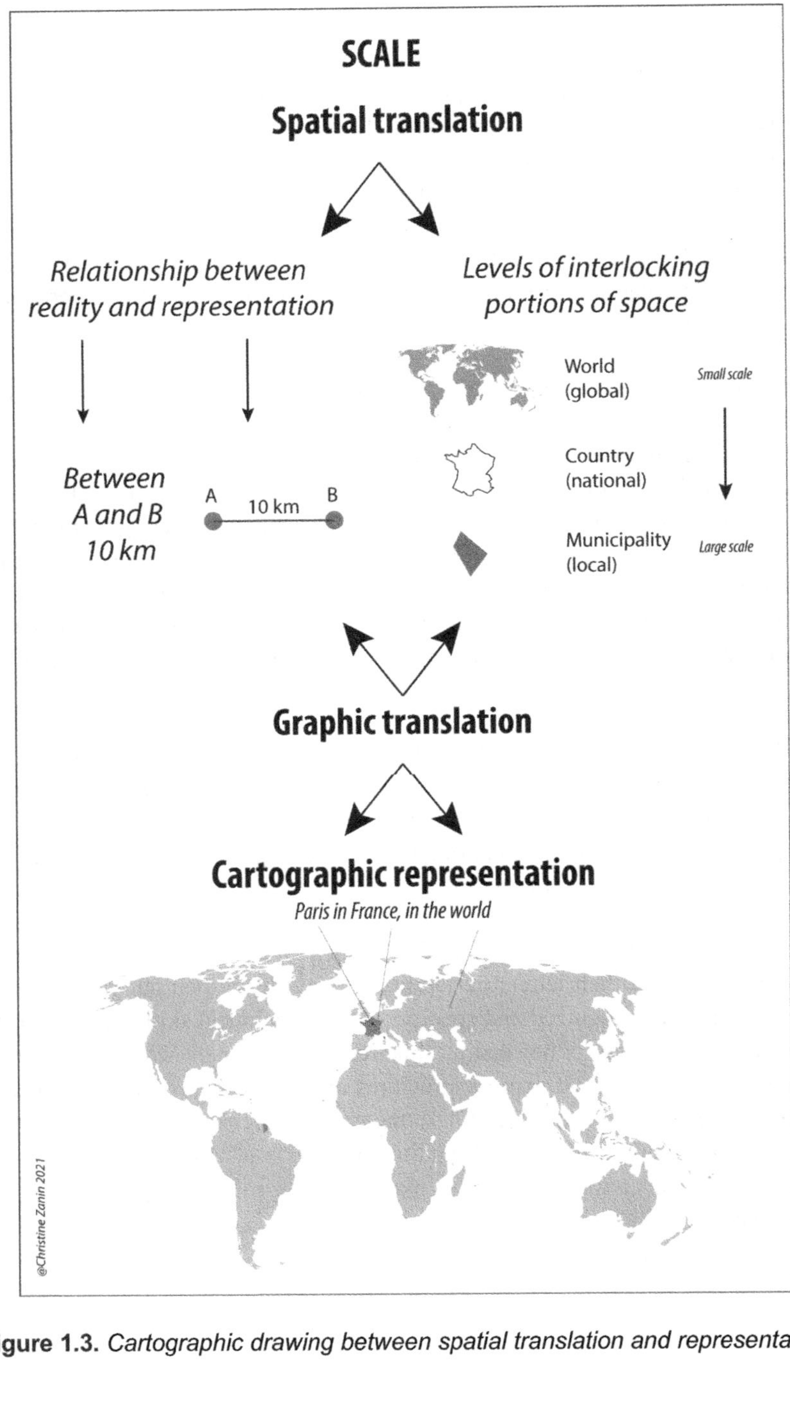

Figure 1.3. *Cartographic drawing between spatial translation and representation*

The Bertin School of Cartography became a DESS in cartography and then a DESS in cartography and GIS to become the Master Carthagéo, which trains geographers-cartographers and cartographic engineers, specialized in spatial analysis and representation of geographic data. We are far from the only graphic ability. The evolution of the title of this Parisian training course expresses the major evolution of the cartographer's work. Their skills are expanding and are moving toward a mastery of "conceptual approaches and methodological expertise of the various formalizations of geographic information". The cartographer's "expertise", in line with technological developments, is doubled as a researcher. It would be necessary to verify whether they are a researcher/cartographer or a researcher/geomatician, which would then of course need to be studied in greater depth and would give rise to numerous debates, but it is nevertheless very clear that their field of action is situated, more than ever, at the intersection of numerous disciplines, the social sciences and the engineering sciences.

The cartographer is therefore no longer quite a craftsman. Yet, here is a state of being that is comfortable to them, an expertise, a technicality, a turn of hand (a *graphos* or *rotring*), almost like an artist who takes their time to elaborate their task. They are not too sure of their identity anymore. Institutionally, they are a geographer, but they will nevertheless go in search of a stronger identity. Without being a jack-of-all-trades, the cartographer becomes nomadic, moving from discipline to discipline, to put their skills as a "spatial translator" (Tobelem Zanin 2013) at the service of numerous themes in the expression of geographical projects and ideas.

1.4. The cartographer, officer of political power

The cartographer-geographer is also an officer of political power, the direct heir of an old tradition, that of the military topographer. These maps were standardized, precise and controlled by the powers that be. They had to be efficient, operational and free of extravagance, representing facts or strategies. The meaning of this idea can be found in a cartographer at the service of land use and planning.

Decision support cartography has developed considerably over time and is no longer a new field: regulatory maps and plans, land use and occupancy maps, maps of territorial aptitudes and constraints, environmental maps and maps of natural and technological risks, etc. It has been adapted to the scientific studies and research carried out almost everywhere in the world, in order to face the problems of land development, urbanization control or environmental preservation. In this context, the place of the map, and more precisely that of spatial representations, is still struggling to establish itself. While all players recognize the importance of cartography, regardless of the scale of the decision, the powerful tools developed in the panoply of geographic information systems tend to make us forget that, from the data to the map, the process must be integrated into the exchanges and discussions between all players in a territorial project (Lardon et al. 2011; Debarbieux and Lardon 2012). The map fulfills an illustrative and exploratory role for specialized research, but it also serves as a communication tool between decision-makers, practitioners and citizens.

Another problem, of purely political nature, is to sort out the purposes of the accumulated data, distinguishing between uses (assessment, control, comparison) and strategic scales (local, regional, national, etc.). In the case of territorial demands, the traditional process is either a precise question addressed to data that will be translated into a map, or an exploratory research on themes for which we try to understand the territorial organizations. Thematic, methodological and conceptual approaches come together around territorial analysis. The construction scheme can then be organized in both directions: from the analytic approach (by questions of elementary description) to a strategic approach (by questions of stakes and helping political decision), the methodological approach that makes it possible to link between the two approaches and to simultaneously reveal new analyses and new policies of spatial organization.

The territory and its analysis remain at the center of this reflective structure. The conception of territories through spatial representations is an integral part of territorial issues (Faludi 1996; Dürh 2007). Maps often appear in the discourse of politicians as a central and essential means of communication between states and the case of transnational thinking. This illustrative aid is all the more valuable as it allows for reflections outside the legal and formal frameworks of each country for the development of spatial policies (Faludi 1996) and often for overcoming language boundaries, as well as those of thought patterns.

However, some difficulties should not be overlooked. Cartographic design and visualization can be difficult to handle in the process of developing territorial planning strategies. The experience of the European research program ESPON, which deals with public policies for spatial planning and development on a supranational scale, is a very good example to be explained (Zanin 2013). Cartographic representations in multiple forms are indeed an integral part of the program's research axes. At the level of projects and governance, they are promoted (websites, cartographic tools, atlases, etc.) and often allow for effective confrontations between players (scientists and political actors or project managers). Halfway between scientific research and political decision, cartographic production is not neutral here: standards of design and presentation of the images produced are introduced and imposed.

The cartography produced is mostly very conventional and "standardized", constrained by short-term objectives of a particular policy, rather than fostering a longer term strategic vision. The graphic representations of the projects are not very original and lack innovation in the way they graphically translate the political message, whether in the traditional publications of the reports, in the specific publications or on the website. The maps produced, as in many political governance bodies, are, in fact, highly standardized and given an extremely strong visual identity. They are maps designed to create consensus. There is no desire for projects to provide original information, and it is accepted that the messages conveyed by the data or the maps are trivial, or already known, and "do not have an impact". Some control through cartographic harmonization, contextualized as a visual identity, is often required.

This standardization of spatial representations is most often done to the detriment of a "carto-diversity" (Zanin 2013), which would be a source of debate, but which would make the realizations more original, and perhaps more interesting and attractive. It is a pity that a certain desire for harmonization blocks the production of cartographic achievements that are abundant in stimulating reflection (Kraak 1998). Each map is controlled in order to deliver a message that is faithful to a pre-established communication strategy, without giving a chance to new forms of representation and thus new perspectives. This is a profound error for the political effectiveness of spatial planning and decision-making strategies. A feedback loop therefore emerges between cartographic innovation and political innovation.

Further cartographic reflection on modes of representation seems necessary (Zanin et al. 2013). However, the political bodies involved in this project often seek more and more standards whose objective is to recognize, at a glance, where the map comes from. The map, through its graphic properties (title block, color), is almost a substitute for the logo. The characteristics of the map's design are so systematic and identical that one could see all these productions as being the same map. The fact that they are effective and achieve their objective of aiding decision-making, as well as acting as a relay between the scientist and the politician is, ultimately, only of secondary importance.

The cartographer "officer" at the service of power can quickly feel frustrated in their missions. This frustration with politics (Robert et al. 2008) ultimately leads the cartographer to think that they are probably a scholar. However, characters remain. The cartographer semiologist and the cartographer artisan support sticking to the idea that has proven itself: one idea per map. Other cartographers refuse this proposal, which they find outdated in the modern geomatic context. It is necessary to find another way and new fields of experimentation.

1.5. The cartographer, crossing borders from geography to social sciences

The cartographer remains attached to the scientific approach that has modified and rethought the geographical approach. They could limit themself to their historical attachment to geography, but their curiosity will make them go beyond traditional academic boundaries. In a period of ever-increasing porosity and interaction between social science disciplines, they will try to transmit the logical order imposed by "graphics" (Bertin 1970) through a precise, constructive and operational visualization of the different stages of a project. Many human and social science disciplines have invested in the spatial field and question the space, the territory and their representations. In their time, Bertin was already aware of it.

The graphic problems encountered during the cartographic realizations go beyond geographical issues and "concern issues of history, ethnography, sociology, psychology, archaeology, architecture, economy, linguistics, politics, demography, geography, medicine" (today, we could add statistics, computer science, communication sciences and media, etc.), "and we notice that graphic solutions are very often the same, whatever the field concerned!" (Bertin 2004, p. 42).

There is therefore something common to all these graphic problems, something totally independent of the scientific fields concerned. This is the origin of the discovery of the graphic expression that is, in France, commonly called "visual variables", and of the specificity of the graphic language of Bertin (1967), which was very close to what Rimbert (1968) proposed the following year. These variables make it possible to graphically represent the levels of measurement of quantitative or qualitative, ordered or differential phenomena on a map. It is true visual grammar elaborated by considering the person who receives the map. There must therefore always be a link between the nature of the data, its geometric implementation (point, linear or zonal) and its representation, "the visual variables serve to guide basic map symbol design" (Krygier 2011). This is the very meaning of effective communication, and not just in geography.

In history, numerous works such as Braudel's on the Mediterranean (1966), Guerreau's (1981) on the mendicant orders of medieval France, or Lepetit's (1988) on cities, present a very varied cartography, which testifies to the influence of Bertin's semiological and graphic work. Admittedly, these are old examples, but they show the importance of semiological influence. By focusing on the analysis of spatial processes and their representation, history as a discipline fully entered the spatial turn of the 1980s.

In the political and sociological sphere, another example is provided by the work of Siegfried (1913), a sociologist, historian and economist who was a precursor of electoral geography, or that of Goguel (1951), who published *Géographie des élections françaises de 1870 à 1951* (Geography of French elections from 1870 to 1951). They show the benefit of using maps to show the political structures of France at the time. In an article published in *Études rurales* (Goguel 1967), they discuss the "great difficulties" of choosing a base map, the lack of weighting when it comes to representing population weights from different spatial contexts, and the problem of constructing proportional circles with percentages. Similarly, they recognize the "unquestionable contribution of the cartographic tool to the work of the political scientist" (Bussi et al. 2019).

In demography and anthropology, Le Bras and Todd's first work (1981), *L'invention de la France*, scanned the French territory to highlight its discrepancies, particularly those related to sociofamily structures. In particular, it allows us to see France through the prism of the departmental division, where social inequalities can be understood by the mobility of rural areas toward urban areas. Their second book, *Le mystère français* (The French Mystery), published in 2013, changes the methodology and mode of representation. The map by department gives way to a smoothed map that shows potential states, rather than definitive states. The structures change, the borders move, and there is a clear opposition between urban centers and their peripheries, more so than between rural and urban areas. Here, the spatial turn is combined with a cartographic turn (Levy 2012), where the revealed territory delivers spatial structures and organizations. Cartography may be a "revelation", but it has mostly become an action.

These examples raise two issues: that of the division of space and the mode of representation. The issue of division seems trivial; the action of dividing a territory into administrative cells seems fairly simple. The data to be represented are collected according to this same division and should therefore not be open to discussion. However, observations and several studies known by the acronym MAUP[5] (Openshaw 1984) clearly show that the division used sometimes has a very strong influence on the reading of the spatial organization of a phenomenon, and it can be considered as a real spatial filter, as Lambert and Zanin (2016) have illustrated (Figure 1.4).

The size of the territorial units has a strong influence on this reading, especially given that certain statistical processes, upstream of their representation, are also

5 The MAUP (modifiable area unit problem) is the principle according to which the division of space used influences the reading of the spatial organization of a phenomenon. The use of a mesh, composed of territorial units of very different sizes, presents non-comparable geographical realities.

influenced by this parameter. It is therefore often preferable to vary the proposed breakdowns in order to highlight the right scientific and political message. For example, it is possible to combine the NUTS 2 (regions in France) and NUTS 3 (departments in France) divisions on the same European map. This combination makes it possible to homogenize the sizes of the grids and thus significantly improve the comparability and perception of the phenomena, without posing any difficulties concerning access to data.

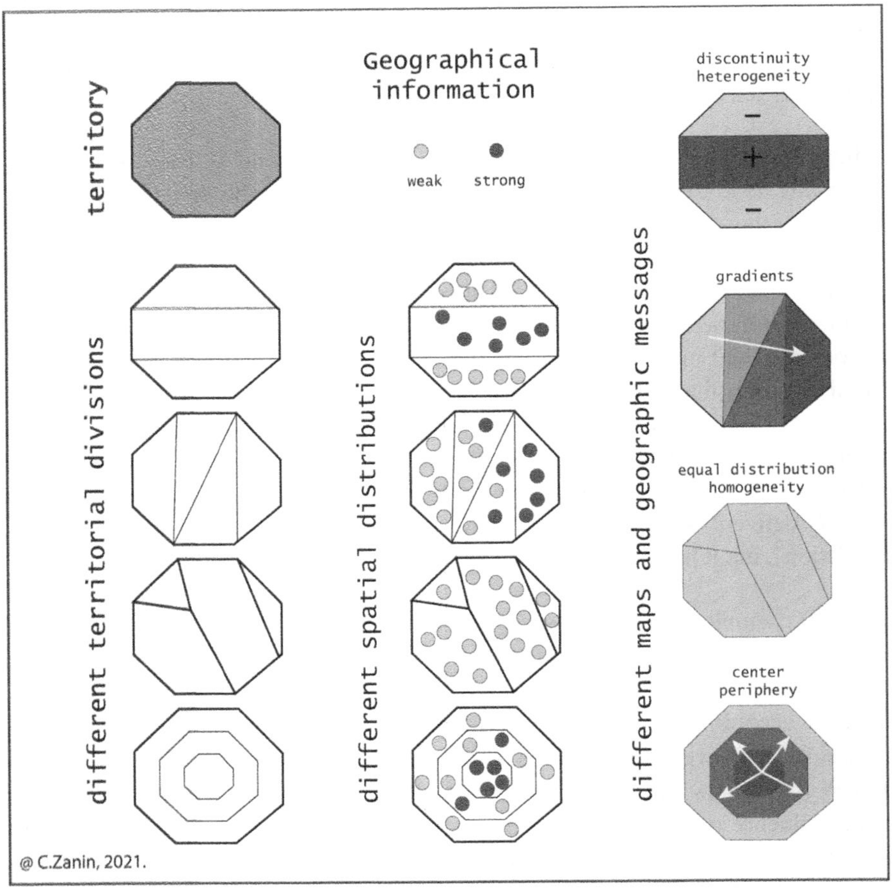

Figure 1.4. *The territorial division as a visual filter (Zanin 2016). For a color version of this figure, see www.iste.co.uk/mericskay/cartographic.zip*

Another important issue is the choice of mode of representation. We can note the great number of maps using flat colors to the detriment of other types of representation such as maps using proportional signs. However, intensity maps are relative maps, presenting information on intensity and order, without considering the size of the spatial units concerned or its link with the gross importance of the phenomenon. In political terms, one could say that these maps are "development" maps, whose purpose is to achieve consensus. In contrast, stock maps highlight absolute weights. These maps are more clearly associated with the notion of power and hierarchy (Figure 1.5).

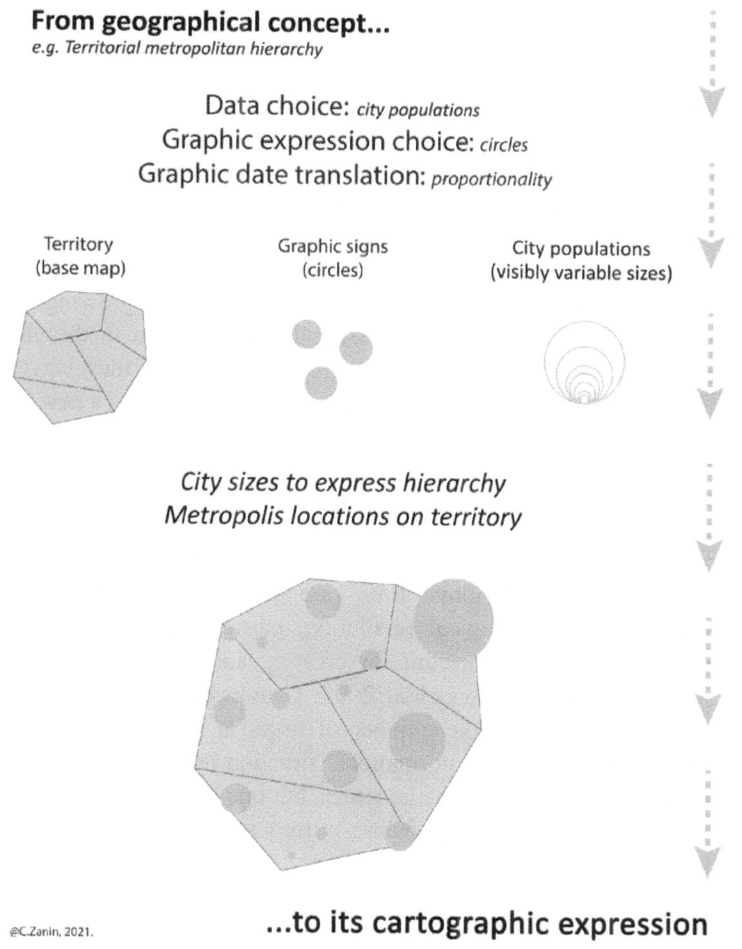

Figure 1.5. *Visualizing a geographic concept (from Zanin 2021). For a color version of this figure, see www.iste.co.uk/mericskay/cartographic.zip*

The representation of the masses introduces a power balance between the territories. The choice between the two is therefore not at all trivial[6]. The explanation of this preponderance can be technical: the producers of maps, who are rather computer scientists or of computer culture, often use a software and technical environment resulting from the GIS, whose standard features propose representing layers of colored polygons. Representation by proportional signs is more complicated. A second reason is certainly more psycho-sociological and comes from the tendency to want a flatter vision of the world in search of a representation that is less polemic.

This fourth cartographer thus acquires a strong identity. They become necessary, and we question them on the choices to be made for a better representation of the spatial aspect. We always work on a small set of data, the data are rare and we get as much as possible, spending time on it and analyzing it in every way. Beyond the issues of the modes of representation, the new informational and technological situation simultaneously questions the identity, competences and missions of the cartographer. The latter must, in fact, face the growing, not to say exponential, increase in data to be processed. They must turn to the engineering sciences; but is that not a betrayal? The cartographer-geographer must waver on their foundations; as ambassador of geography, they are obliged to cross a new border.

1.6. The cartographer, an expert in geographic data

The problem of the fifth facet of the cartographer is not moving from one discipline to another, but to move from rare data to abundant and extremely diversified data. The technological revolution, particularly cartography's digital conversion, has changed the medium for creating and disseminating geographic information, and thus maps (Goodchild 2007; Cauvin et al. 2008). Databases are proliferating and opening up, the number of users is increasing and geolocated content is flourishing. This new informational situation leads us to think and design new analyses and explorations of spatial issues, but one difficulty persists: analyzing is one thing, but being able to translate it into spatial terms is another[7]. The confrontation of data and their relation to each other remains a major issue of the cartographer's work. However, the overabundance of geographic information, its heterogeneity, as well as its sharing, radically modify the status of the map and the skills to be implemented for its design and distribution.

6 See Chapter 3, which reviews the issues of representation modes in thematic cartography.

7 See Chapters 4 and 6 on graphic semiology issues on the Geoweb.

In cartography, everything either starts with a data set, leading to its representation, or with a working question/hypothesis, leading to the research or production of data. The capacity of information systems to store data, as well as new mobile and Web technologies that set up new data acquisition devices, contribute to the accumulation of geographic data. Due to the diversity and evolution of tools and methods of production, as well as processing, these data are heterogeneous, imperfect, multi-level and multi-dimensional. They are produced at different spatial and temporal scales and, moreover, with different objectives. The stored geographic information integrates a temporal aspect (the data are collected at a precise date and duration), which can be more easily considered in cartographic terms, and makes it possible to follow territorial transformations and dynamics: expansion of cities, evolution of growth and demographic indicators, modifications of landscapes and land use modalities, spatial dynamics of environmental phenomena (pollution, climate, vegetation, etc.), movements of individuals and phenomena in space according to time, etc.

Gigantic spatial databases are accumulating and inducing new analyses and explorations of issues, in addition to new dynamics through the introduction of "continuously" collected data (Kitchin 2013): sociospatial or spatioenvironmental phenomena, study of individual behaviors, sociological, regional and urban dynamics, relations between societies and environment, etc. Big Data is entering the world of geography by disrupting not only quantitative approaches but all issues of visualization of these analyses. Referred to as "major perspectives of the sciences" (Marx 2013), the "generalization of geolocation tools (by GPS, GSM, Wi-Fi or even IP address) leads to an explosion in the availability of spatially referenced data" (Audard et al. 2014, p. 1).

These databases have never been so numerous. They continue to include geometric or semantic data, but are now supplemented by individual databases. Thanks to applications that locate the position of individuals or objects, track their movements and allow trajectory analysis, it is possible to produce what is called "voluntary" geographic information, of which OpenStreetMap is one of the best represented examples (Figure 1.6). The geographic web (Geoweb) has become one of the means for storing, publishing, sharing and disseminating geographic data. Associated with the development of Volunteered Geographical Information (VGI), where each individual can become a contributor in the process of producing geographic or geolocated information (Goodchild 2007), cartography is evolving and transforming[8]. A new revolution is also underway, that of Open Data.

8 See Chapters 2 and 5 on the impacts of these new dynamics on the cartographic communication process.

Administrations and local authorities have experienced this revolution first hand, and have yet to figure out the path to "transparent and adapted visual communication".

Thus, the mass of new data is constantly increasing and, in turn, the difficulty in analyzing them, according to all the thematic aspects they offer, is growing. Many processes remain poor, considering the possibilities. All these databases, that are both rich and heterogeneous, are rarely linked together. The confrontation of data and their connection remains a major issue for their cartographic representation: how does one express data with different accuracy and coming from multiple producers, all on the same map? How can we have confidence in data that are manipulated and produced under very different conditions, with qualities that are just as different and that are not always compatible? How can we avoid making errors in interpretation and analysis?

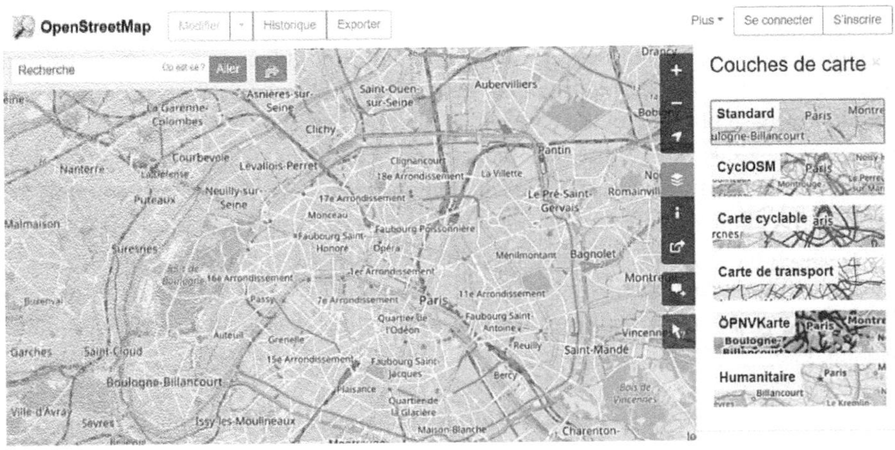

Figure 1.6. *OSM, a cartographic project for free and complete geographic information (source: OpenStreetMap). For a color version of this figure, see www.iste.co.uk/mericskay/cartographic.zip*

With the rise of digital technology, the raw material of the cartographer is no longer the same, nor is the technical environment. Imagine a medieval stonecutter in front of an American building: everything has changed. The cartographer craftsman as such disappears in favor of a character who must now accept moving to a higher, industrial-type stage and find other ways of working, because the tools, as well as the cartographic materials, have changed. The cartographic approach remains the same; the logic of the workshop where the cartographer seeks to produce the best possible graphic message, the issue of representation, is coupled with a perilous and complex practice, that is of the acquisition, structuring, storage, processing and dissemination of geographic information accompanied by the learning of new tools.

This question is all the more crucial as it deals with issues of territorial analysis and decision support and also complicated by a political issue. However, the large masses of data of all kinds do not make it possible to assemble and solve everything. On the contrary, they generate new problems and new questions for the cartographer, who must maintain a strong link with statistics, modeling, spatial analysis and dynamic visualization. The cartographer must accept the fact that they are to merge into a new profession, into new logics and new skills: it is a great challenge. The production of geographic information and maps is no longer reserved for professionals alone. This constitutes a real culture shock for the cartographer craftsman (Palsky 2013).

A cartographer who discovers computerization of data processing and its representation has not become a computer scientist. They are no longer a semiologist or a craftsman. Why is it still necessary to fight on the semiological front? Have the digital world, geo-visualization, dynamic cartography, participatory, interactive and animated cartography made us lose the notions (not imposed but theorized) presented by Bertin and many others? Is it useful or futile to persist in representing stocks in the form of proportional circles, and not in flat areas of color?

Certainly, with the plethora of contemporary tools (GIS, webmapping, programming language, etc.), it is easy to mobilize any semiotic form or range of colors, but should the notion of a geographically oriented and adapted message dissolve into a graphic overload, without an effective representation of the updated spatial structures? The notion of information density transmitted by map "from the sender to the receiver" (Moles 1964)[9] must remain essential in cartographic design. The increasing computerization of production can lead to the cartographer losing skills: no longer a craftsman, just a technician at the service of expert systems that are supposedly capable of applying the rules of semiology, without error. In short, a profession has "disappeared", like blacksmiths or saddlers, but others have emerged, like the geomatician or the geo-data scientist. However, the cartographer-geographer is tenacious. They affirm loud and clear that they will always remain a communicator, and that they have various ideas to go from the craftsman stage to the industrial stage: they propose a multi-map!

9 According to Abraham Moles, one of the challenges of the cartographer is to lower the "information density" of the map in order to achieve a balance between redundancy and originality of information. They thus suggest "providing less than ten binary elements of effective originality per second".

1.7. The art of multiplying maps to communicate better

Through technological, methodological and cultural developments, the cartographer has acquired a culture of data manipulation and visualization, where the map is only one of several modes of representation, with the emergence of additional facets referring to new fields such as data journalism or data visualization. One thing is certain: despite this evolution of practices, the cartographer is, and remains above all, a communicator, an artist of transformation of ideas into an attractive visualization. In order to reach their objective and attract the reader, the cartographer must play on spatial and temporal scales by proposing an alluring scenario and attractive settings.

The deliberate manipulation of the scales aims to suggest other spaces of reflection and action both on the territory and the social side. Choosing a scale is not trivial. This choice consists of selecting a level of information that is relevant to the level of territorial and social organization to be studied. Drawing a plan to scale does not fundamentally amount to establishing a relationship of reduction between reality and its representation, but it is, above all, selecting one dimension and renouncing the others; it is emphasizing the choice and the intention that it implies. Two figures on the same space, that are different, do not reveal the organization of the space in the same way. The current cartographic instrumentation, especially online, makes it possible to approach and solve these issues by proposing interactivity, dynamism and animation in the representation.

Interactive access to geographic information meets a growing demand from territorial players, who wish to visualize their territory, compare it to others, situate it in different contexts, and decide with the "best" maps. Interactivity allows this by approaching the field of differentiated views[10]: display, search for information and change of scale. Cartographic representations remain at the heart of these devices and allow to propose a story that is not yet been written or defined. The user/reader has the means to illuminate his or her own data set to construct his or her own understanding of the phenomenon. The map becomes the interface that allows the database to be examined, explored and analyzed. The objective is not only to locate or map, but to explore the links between different indicators and space in order to build a scenario.

This is good and even better, but it is a problem for the user, who risks getting lost in the proliferation of infinite cartographic possibilities. Once again, to get through this maze of images, we need a well-constructed scenario that tells a story; one that must be captivating enough to reach your readership.

---

10 See Chapter 7, which returns to the question of interactivity in geo-visualization in more detail.

The work of multi-representation cartography[11] (Zanin and Lambert 2012) is a very effective avenue for anyone who wants to build such a scenario[12]. These types of applications are based on the multiscalar comparison of cartographic representations of territories. Three types of comparison are possible: between mesh levels, between representation modes and according to the scales of analysis. Several atlases or interactive websites propose the principle of "spatial wandering" (Debarbieux 2002), which is made possible by the numerous views, types of representation and levels of analysis of the same space and theme (Figure 1.7).

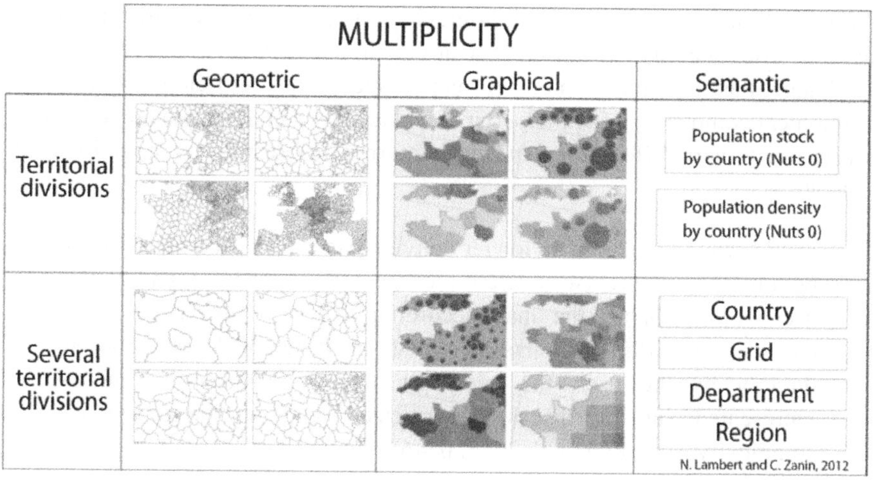

Figure 1.7. *Geometric, graphical and semantic multiplicity (Zanin and Lambert 2012). For a color version of this figure, see www.iste.co.uk/mericskay/cartographic.zip*

Viewed from this angle, multi-representation makes it possible to overcome two well-known cartographic problems, that is, the superimposition of information, which can certainly also be resolved by mathematical processing or adapted modeling upstream of the representation (Cauvin et al. 2008), and the superimposition of excessive number of graphic signs; the second problem is related to the need to adapt the representation to the intended message, to the audience of the map and to the context of its production.

11 Multi-representation is understood as a multiple cartographic realization; in other words, the representation of the same phenomenon using different modes of representation (different map backgrounds, different scales, different calculations from the same data and different styles).

12 This perspective of multi-representation is implemented in Chapters 3 (around the theme of world population) and 4 (on the example of global container flows).

Although some representations may seem contradictory, they are instead complementary in a panel of possible images. It is about offering a cartographic panel where each map can enrich reflection, analysis and decision making. The notions of contextualization and multiple representations become key notions in cartographic exploration environments. Conceived in this way, multi-representation no longer presents the map as a tool of power or domination, influencing and directing the reader toward the uniqueness of a solution, but as an instrument or means of reflection.

Based on this principle, a new cartographic object was created, the multi-map: a map that varies according to different scales and representation modes, but that does not change the data represented. A map alone is not sufficient to reflect the geographical complexity of reality, so it is replaced by a coherent collection, leaving the user free to vary different parameters, capturing different facets of reality as they or she wishes.

> In this way, the map is no longer delivered as is, but is co-constructed both by the designer (who determines the possible choices) and by the user (who shapes the map according to his or her needs). By considering different points of view, at different levels, chosen directly by the user, the geographical phenomenon thus multi-represented can be discussed, criticized, and questioned in its complexity. (Lambert et al. 2015, p. 46)

The multi-map solves a problem: the reader is no longer forced to choose a single solution from the infinite number of possible solutions. However, we then fall into dismay for a reader who is summoned to choose between multiple visions of the same reality. What is welcomed by the scientist (freedom of choice and representation) is often rejected by the politician (in search of standards and simple solutions). It must then be admitted that the diversification and multiplication of maps mixes time, representation spaces, players and techniques. It is essential to be coherent by proposing a presentation that gives full effectiveness to the scenario put forward. One of the solutions to be explored is the implementation of a cartographic framework, which would have a place in this necessary presentation in order to develop or reinforce these new tools of territorial understanding. The "multi-map" (Zanin and Lambert 2012) would then be associated with its commentary, equally numerous, and would become a very effective operational element.

Even though multi-representation offers many different maps, not all of them have the same value. Some provide a lot of information and are relevant in all circumstances, others only show marginal information. The task that then emerges is to design a cartographic scenario by integrating commentary alongside the set of maps offered. Instead of letting the user "pick and choose" the maps they want to see, they are displayed in a rational order, by clicking on links or navigating from page to page, as one progresses in the reading of the text in a story-telling manner. The maps are thus presented successively, according to specific locations and levels of zoom or points of view, like so many images supporting the commentary carried by the text. The cartographic tool is transformed into a real scientific exploration toward an editorial communication that is adapted and oriented toward the largest audience (Figure 1.8).

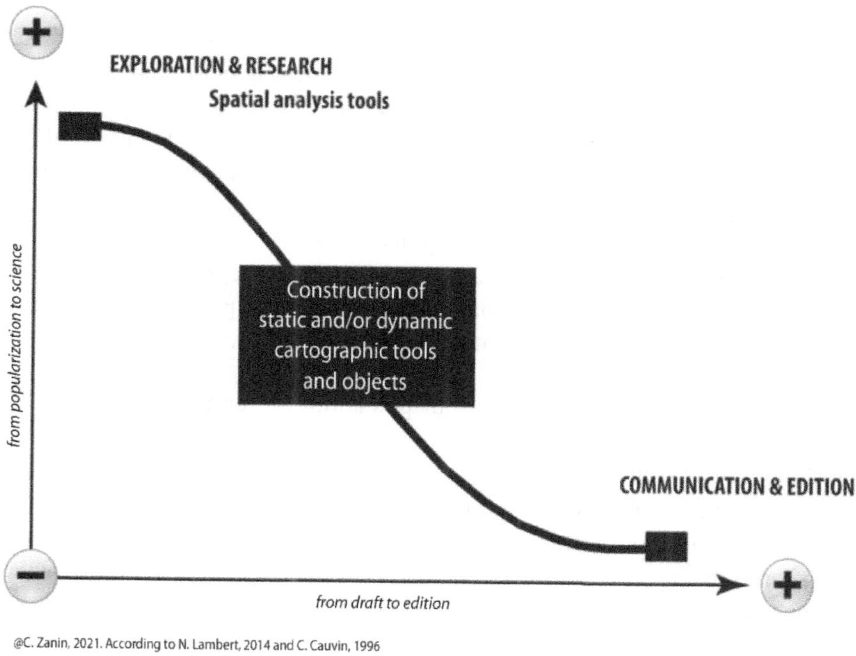

Figure 1.8. *The cartographer's place in geographical communication*

This cartographic project seems to be a real innovation, not only in the world of maps but also in the field of land-use planning policies. A dialogue between cartographers and readers, or between inhabitants and territorial players, can then be set up. This dialogue does not rely on a single participatory representation (Palsky 2013), but is developed around and about the map, in order to allow the construction of a variety of cartographic proposals, adapted to the reader. Several interactive tools have been built according to this principle over the past 15 years. Among them, the

HyperAtlas[13] and the European atlas AIRE[14] have taken full advantage of this solution. However, no one cartographic tool is better than the other. For example, although dynamic or animated cartography is a good option for dialogue within the framework of the implementation of territorial planning options, or simply to follow the evolution of a spatial phenomenon, static cartography, whether it is multiplied or not, is an equally relevant option for popularization communication. It is not a question of opposing certain cartographic options but of adapting to the project.

1.8. Conclusion

It is no longer necessary to demonstrate the social and societal utility of the cartographer. Their technicality, their rigor, their artistic sense and their way of reinventing the (spatial) visualization of phenomena make them a driving force of an applied and ever more operational geography. The different facets of their profession, developed in this chapter, make them a character capable of proposing analytical solutions and setting up pragmatic approaches to the representation of space in diversified contexts by mobilizing data, methods and tools in the right way. For this particular type of image maker, the question of representation is coupled with a question of the accuracy and/or validity of the image proposed by the map, as well as its congruence to the represented or perceived reality.

As Tiberghien (2007) points out, "a map is a tool for representing reality, but it only represents it, today as in the past, in a partial and biased manner." There is no such thing as an optimal map (MacEachren 2004) or an objective representation. The cartographer proposes their way of approaching space by constructing a cartographic image that must only be considered as a version or an interpretation of the territorial reality to which it refers, and always according to the cartographer's intentions, the target audience and the context in which it is placed. The map is not a "mirror" of the territorial reality but an interpretation, an "optional" construction of the territory.

The notion of an option refers to the choices that the cartographer is confronted with at each stage of the construction of a map. It can guide decisions and become a

13 The HyperAtlas software is a multiscalar measurement and cartographic tool for territorial inequalities. Developed by the HyperCarte research group, it allows for a multiscalar territorial analysis, which consists of considering that the meaning of an indicator is always dependent on a territorial context that serves as a reference.

14 AIRE is the interactive atlas of European regions. Designed at the end of 2006, at a time when there were few online maps, it allows the visualization of sociospatial phenomena at different scales because of several methods of representation and spatial analysis.

real power issue through the proposal of "competing representations, between competing knowledge and values" (Besse 2008). This raises the issue of both the usefulness and the effectiveness of this document. The notion of a cartographic version is more a matter of visualization than of selecting the mode of representation. It implies choices of a "set of graphic procedures, intended to facilitate the interpretation of data sets" (Guichard 2019). The "tricks" used for effective visualization refer to esthetic practices and innovations. The cartographer tries to attract by proposing alluring colors and shapes for a targeted audience, or original or even innovative representation modes. They refer to everyone's visual perception, which makes the map fragile, but provides it with a force of attraction.

The question that naturally arises is to understand whether, in order to be attractive, or even effective, a map must be qualified as "beautiful". Worringer (1941) indicated that estheticism (especially in art) is the expression of a particular intention, and what is known as "scientific esthetics" is basically just a psychological interpretation of a style. Beauty is not the only goal of the esthetic. In this sense, the artistic intentions of the cartographer must be questioned. These intentions are not indisputable; it is necessary to consider what the cartographer knew how to do, but also what they wanted to do, what the image seeks to express. The graphic signs treated according to certain rules, traced with various graphic modes, affixed on empty or full places with embellishments or highlights, constitute so many ways and so much symbolism, which are superimposed on the semantics.

Another aspect is set up, that is, of the link that must exist between effectiveness, efficiency and satisfaction (Christophe 2018). If cartographic effectiveness makes it possible to approach cartographic realization as expected by users according to their needs, then efficiency dictates that it must be possible with a lesser effort, making it possible to gain the satisfaction of understanding the result. In order to be effective, cartographic production requires a careful balance between the simplicity of the representation modes, the sensible use of graphic semiology and the organization of the modes of representation and the image. The map can and must tell a story, even a simple one, through the construction of legible and effective images. This becomes effective if the link between visual perception and thematic understanding is made as quickly as possible. Bertin (1967) defines effectiveness as the reader's ability to "obtain a correct and complete answer to a given question". However, the effectiveness of a map is not just measured by a reduced comprehension time and is not absolute. It depends on perceptual habits and on the education of the reader (Palsky 2017). As Harley (1991) points out, cartographic efficiency cannot be reduced to "the application of inflexible drawing rules" and a single graphic semiology. Visual expertise must be coupled with a very clear understanding of the context and purpose of production: what cartographic approach? For what map? For what message? For what purpose?

At a time when maps are omnipresent in the press, on the web and social networks, the cartographer should not take for granted the fact that the representations they propose are usable and understandable by everyone. Moreover, the representation of the territory is no longer the cartographer's own playing field, but is part of a multitude of disciplines and fields of practice where cartography, image processing, mathematics, computer science, computer graphics, data visualization, geo-visualization, data journalism, etc., are intermingled.

Between an option and cartographic version, analysis and communication, scientific object and artistic image, and reflection and decision, cartographic design refers to multiple aspects that mobilize all facets of the cartographer. Despite all the pitfalls, the cartographer possesses a powerful, innovative and constantly changing tool that allows (and requires) them to constantly renew and reposition themself as the "guarantor" of cartographic communication. The facets of the cartographer are just waiting to multiply, to develop according to technological and methodological evolutions, but above all, to adhere to the needs of representation, which must continue to effectively carry a discourse, whether it be scientific or political, or to simply send a desired message. The power of the cartographer is considerable, as are their responsibilities, hence the importance of continuously thinking about the most appropriate representations and communication for each subject.

1.9. References

Andrienko, N. and Andrienko, G. (1997). Intelligent cartographic visualization for supporting data exploration in the IRIS system. *Programming and Computer Software*, 23(5), 268–282.

Audard, F., Carpentier, S., Oliveau, S. (2014). Les "big data" sont-elles l'avenir de la géographie ? In *20e Biennale de géographie d'Avignon. Géopoint 2014 Controverses et géographies.* Avignon.

Bahoken, F., Grasland, C., Zanin, C. (2016). D'une cartographie de flux à une cartographie du mouvement. Aspects sémiologiques. *Cartes et Géomatique*, 229/230, 65–74.

Barthes, R. (1985). *L'Aventure sémiologique*. Le Seuil, Paris.

Bertin, J. (1967). *Sémiologie graphique. Les diagrammes, les réseaux, les cartes*. Gauthier-Villars-Mouton, Paris.

Bertin, J. (1970). La Graphique. *Communications*, 15, 169–185.

Bertin, J. (1977). *La graphique et le traitement graphique de l'information*. Flammarion, Paris.

Bertin, J. (2004). Aux origines de la sémiologie graphique. *Cartes et Géomatique*, 181, 42–44.

Besse, J.M. (2010). De la géographie comme dimension de la culture. *L'Espace géographique*, 39(3), 193–196.

Bianchin, A. (2012). Actualité de l'approche de Jacques Bertin dans l'enseignement de la cartographie. *Cartes & Géomatique*, 212, 13–20.

Boyd, D. and Crawford, K. (2012). Critical questions for big data. *Information, Communication and Society*, 15(5), 662–679.

Braudel, F. (1966). *La Méditerranée et le monde méditerranéen à l'époque de Philippe II*. Armand Colin, Paris.

Brunet, R. (1987). *La carte : mode d'emploi*. Fayard-Reclus, Montpellier.

Bussi, M., Voilliot, C., Le Digol, C. (2019). *Le tableau politique de la France de l'ouest d'André Siegfried. 100 ans après. Héritages et postérités*. Presses universitaires de Rennes, Rennes.

Cauvin, C. (2016). La sémiotique en cartographie au XXIe siècle. *Cartes et Géomatique*, 229/230, 17–31.

Cauvin, C., Escobar, F., Serradj, A. (2008). *Cartographie thématique. Des voies nouvelles à explorer*. Hermès-Lavoisier, Paris.

Christophe, S. (2018). Géovisualisation(s). Report, Géoviz, Aussois.

De Martonne, E. (1934). *Annales de l'Université de Paris*. Librairie du Recueil Sirey, Paris.

Debarbieux, B. (2002). Figures (géo)graphiques et prospective. Les cartes, schémas et modèles au service du projet et de la prospective territoriale. In *Ces territorialités qui se dessinent*, Debarbieux, B., Vanier, M. (eds). Éditions de l'Aube/Datar, Paris.

Debarbieux, B. and Lardon, S. (2003). *Les figures du projet éditorial*. Éditions de l'Aube/ Datar, Paris.

Dürh, S. (2007). *The Visual Language of Spatial Planning. Exploring Cartographic Representations for Spatial Planning in Europe*. Routledge, London.

Faludi, A. (1996). Framing with images. *Environment and Planning B*, 23(1), 93–108.

Goguel, F. (1951). *Géographie des élections françaises de 1870 à 1951*. Presses de Sciences Po, Paris.

Goguel, F. (1967). L'Atlas de la France rurale à la lumière de la science politique. *Études rurales*, 27, 55–58.

Goodchild, M., Yuan, M., Cova, T. (2007). Towards a general theory of geographic representation in GIS. *International Journal of Geographical Information Science*, 21(3), 239–260.

Guerreau, A. (1981). Analyse factorielle et analyses statistiques classiques : le cas des ordres mendiants dans la France médiévale. *Annales. Économies, Sociétés, Civilisations*, 5, 869–912.

Guichard, E. (2019). Cartographie et visualisation. *Annales des Mines. Responsabilité et environnement*, 94, 38–41.

Guilhot, N. (2005). Histoire d'une parenthèse cartographique. Les Alpes du nord dans la cartographie topographique française aux 19e et 20e siècles. PhD Thesis, Université Lyon 2, Lyon.

Harley, J.B. (1989). Deconstructing the map. *Cartographica. The International Journal for Geographic Information and Geovisualization*, 26(2), 1–20.

Jacob, C. (1992). *L'empire des cartes. Approche théorique de la cartographie à travers l'histoire.* Albin Michel, Paris.

Jacob, C. (1996). Toward a cultural history of cartography. *Imago Mundi*, 48, 191–198.

Jacob, R.J.K. and Karn, K.S. (2003). Eye tracking in human computer interaction and usability research: Ready to deliver the promises. In *The Mind's Eye: Cognitive and Applied Aspects of Eye Movement Research*, Hyönä, J., Radach, R., Deubel, H. (eds). Elsevier, Amsterdam.

Jégou, L. (2016). L'imagination esthétique dans la conception graphique des cartes : proposition de typologie illustrée. *Cartes & Géomatique*, 229, 231–248.

Kitchin, R. (2013). Big data and human geography: Opportunities, challenges and risks. *Dialogues in Human Geography*, 3(3), 262–267.

Kraak, M.J. (1998). The cartographic visualisation process: From presentation to exploration. *The Cartographic Journal*, 35(1), 11–15.

Kraak, M.J. and Ormeling, F.J. (2009). *Cartography: Visualization of Spatial Data.* Prentice Hall, Hoboken.

Krygier, J. and Wood, D. (2011). This is not the world. In *Rethinking Maps: New Frontiers in Cartographic Theory*, Dodge, M., Kitchin, R., Perkins, C. (eds). Routledge, London.

Lambert, N. and Zanin, C. (2016). *Manuel de cartographie.* Armand Colin, Paris.

Lambert, N., Ysebaert, R., Zanin, C. (2015). Comprendre le territoire européen. Deux outils cartographiques à l'appui de la réflexion en aménagement du territoire. *L'Information géographique*, 79, 45–54.

Lardon, S., Maurel, P., Piveteau, V. (2001). *Représentations spatiales et développement territorial.* Hermès, Paris.

Le Bras, H. and Todd, E. (1981). *L'invention de la France. Atlas anthropologique et politique.* Gallimard, Paris.

Le Bras, H. and Todd, E. (2013). *Le mystère français*. Le Seuil, Paris.

Lepetit, B. (1988). *Les Villes dans la France moderne (1740–1840)*. Albin Michel, Paris.

Lévy, J. (1998). *Le tournant géographique. Penser l'espace pour lire le monde*. Belin, Paris.

Lévy, J. (2012). Cartographic turn? Bridging the gap between sciences and technologies of the inhabited space [Online]. Available at: https://www.espacestemps.net/articles/a-cartographic-turn/ [Accessed 15 May 2013].

MacEachren, A.M. (2004). *How Maps Work. Representation, Visualization, and Design*. Guilford Press, New York.

MacEachren, A.M. and Taylor, D.R.F. (1994). *Visualization in Modern Cartography*. Pergamon, Oxford.

Maggi, S., Fabrikant, S.I., Imbert, J.P., Hurter, C. (2015). How do display design and user characteristics matter in animations? An empirical study with air traffic control displays. *International Journal for Geographic Information and Geovisualization*, 51(1), 25–37.

Marx, V. (2013). The big challenges of big data. *Nature*, 498, 255–260.

Moles, A. (1964). Théorie de l'information et message cartographique. *Science et enseignement des sciences*, 32, 11–16.

Openshaw, S. (1984). The modifiable areal unit problem. *CATMOG*, 38.

Palsky, G. (2003). Éléments pour une histoire de la sémiologie graphique avant Jacques Bertin. *Bulletin de l'Association des géographes français*, 183–194.

Palsky, G. (2011). Map design vs sémiologie graphique. Reflection on two currents of cartographic theory. In *25e Congrès international de cartographie (ICC)*. Paris.

Palsky, G. (2013). Cartographie participative, cartographie indisciplinée. *L'Information géographique*, 4(4), 10–25.

Palsky, G. (2017). La sémiologie graphique de Jacques Bertin a cinquante ans ! [Online]. Available at: https://visionscarto.net/la-semiologie-graphique-a-50-ans [Accessed 15 May 2021].

Palsky, G. and Robic, M.C. (1997). Aux sources de la sémiologie graphique. In *Colloque 147, 30 ans de sémiologie graphique*. Paris.

Robert, P. and Souchier, E. (2008). La carte, un média entre sémiotique et politique. La carte au rivage des sic. *Communication & Langages*, 158, 25–29.

Robic, M.C. (1997). Une école pour des universitaires placés aux marges de l'expertise : les années trente et la cartographie géographique. *Cybergeo: European Journal of Geography* [Online]. Available at: http://journals.openedition.org/cybergeo/1643 [Accessed 15 May 2021].

Rimbert, S. (1968). *Leçons de cartographie thématique*. DL, Paris.

Robinson, A. (1952). *The Look of Maps*. University of Wisconsin Press, Madison.

Siegfried, A. (1913). *Tableau politique de la France de l'Ouest sous la Troisième République*. Armand Colin, Paris.

Slocum, T., McMaster, R., Kessler, R., Howard, H. (2005). *Thematic Cartography and Geographic Visualization*. Prentice Hall, Hoboken.

Tiberghien, G. (2007). *Finis terrae : imaginaires et imaginations cartographiques*. Bayard, Paris.

Tobelem Zanin, C. (2013). Cartographie et analyse territoriale. PhD Thesis, University of Paris Diderot, Paris.

Woringer, W. (1941). *L'Art gothique*. Gallimard, Paris.

Zanin, C. and Ben Rebah, M. (2011). Les variables visuelles de la cartographie dynamique : analyse, critique et propositions. In *Dixièmes rencontres de Théo Quant*. Besançon.

Zanin, C. and Lambert, N. (2012). La multireprésentation cartographique. Exemple de l'Atlas interactif des régions européennes. *Cartes & Géomatique*, 213, 39–64.

Zanin, C., Lambert, N., Davoine, P.A., Mathian, H., Saint-Marc, C. (2013). Espon cartographic language. Mapping Guide. Report, ESPON ECL Project, UMS Riate, CNRS, Paris.

2

Cartography as a Communication Tool: Thoughts on Models

Laurent JÉGOU
LISST, University of Toulouse-Jean Jaurès, France

2.1. Introduction

The model of communication in cartography, which is our starting point, refers to several proposals that have evolved over time. One of the best known was formalized by Robinson in 1960, adapting the model of Shannon and Weaver (1949). It was then introduced in France by Moles in 1964, and taken up by Bertin (1967, 1978)[1].

To model how the map "works", as MacEachren (1995) puts it, is to not only study the way in which the map is designed and produced but also used, as well as its effectiveness in these uses. More than a simplification, modeling is seen here as a means to break down, simplify and then analyze this functioning in order to be able to reflect on its articulation, evolution, applicability to our practices and to discover ways of linking it with different approaches outside cartography or geography. This work therefore has a reflective intent, about our research, education and, more generally, the use of maps in geography.

1 For a detailed presentation of the history of the communication model in cartography, see the article by Plantain (2014).

Cartographic Communication,
coordinated by Boris MERICSKAY.

The model of communication, which is evoked in most cartography textbooks, has structured the theoretical developments of the discipline to the point that epistemological research on the subject cannot be free of it. The prevalence of this model in education and research could lead one to believe that certain historical functions of the map still seem to be relevant, even though the contexts in which maps are produced and used are changing considerably.

Given that it was developed in a rather dated context, is this model still relevant today for understanding the process of cartographic communication, especially with the development of other modes of design, production and use of maps, as well as other media and technologies? Can it still serve as a supporting resource for critiquing cartographic functioning, considered from the point of view of communication?

We hypothesize here that this model, which is widely recognized and mobilized, but ultimately rather simplistic and dated, can be the starting point for a contemporary reflection on cartographic communication. The idea is to question it, challenge it, develop it, enrich it, etc., in order to reflect on the role of the map in a scientific and technical context of deeply modified uses, as well as to put the forms, challenges and possibilities of innovation in cartography into perspective.

From the "traditional" model of communication, we will look for elements of semiotics, as well as information and communication sciences, while adopting a rather broad outlook. This is an opportunity to take a critical step back on the way we consider the object of cartography as a discipline. Here, we will only mention some cartography trends in order to connect them to the different variations of the model. Our intention is not to provide a complete description or an exhaustive state of the art, but rather to propose a guide to navigate between these various approaches.

The aim of this chapter is thus to critically examine the model of communication, according to a heuristic method, by modifying and progressively developing its form, starting from visual representations in the form of graphic models, which will be progressively made more complex. The first part will focus more specifically on the functioning of the map as a means of communication, using approaches from semiotics and communication sciences, and seeking to integrate changes in media, techniques and practices. The second part will take into account the duality of the map: a practical tool, but also a graphic expression that acts through its esthetic properties by playing on a certain quality of aspect and sensations. We will conclude with a more synthetic cross-sectional look at innovation's place in cartography in these various developments of the model.

2.2. Are maps a simple means of communication?

2.2.1. *The communication model in cartography*

The diagram presented by Moles in a famous 1964 article is directly inspired by the standard model of information theory: a sender transmits information to a receiver via a message by means of a channel (Figure 2.1). This simplified description of its functioning has made it possible to think about cartography from a communication approach, and to introduce the problems of transmission quality and cartographic language at a pivotal moment in the history of science. This diagram describes a quite dated concrete process linked to the context of the period. This model considers the map as a physical object, produced by specialists practicing complex methods, to be read and used by "readers", map users who use it as a source of information to guide their actions.

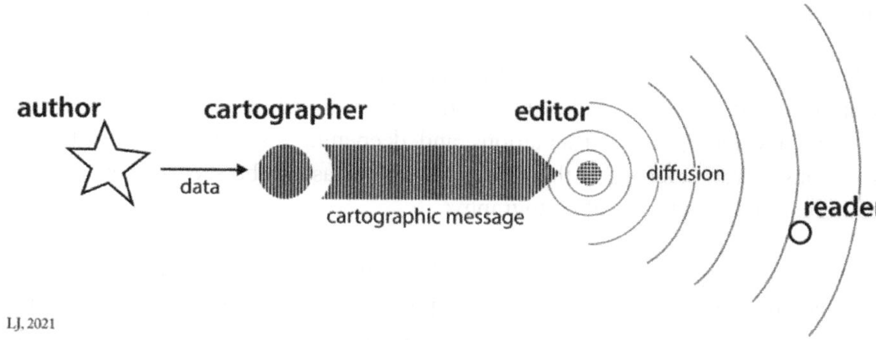

Figure 2.1. *Cartographic information process (according to Moles 1964)*

In this line of thought, Moles complexifies the information theory model by adapting it to publication cartography (in other words, on paper), in an essentially scientific and educational context. Thus, the production of the map may involve three different agents: the author, who may be different from the person in charge of the realization, the cartographer themself and lastly the publisher ensuring the dissemination, as in the field of literary publication. We can note here the undulatory graphic metaphor of diffusion used by Moles, like in the case of a radio broadcast. The users, like the listeners of a radio program, "receive", mobilize and interpret the map among a field of indeterminate possibilities. The map is proposed here as a physical object, whose reception (time, duration, conditions) and interpretation by its audience (understanding, subsequent use, etc.) is not controlled. The audience may then seem rather passive. Once produced, the map is finished and continues its existence without any possible intervention from its producers.

2.2.2. *First extensions of the model*

In 2001, Sui and Goodchild noted that research in cartography, like in geomatics, was focused on questions of transmission, technique and "grammar" (syntax) in communication, forgetting those related to the reception and use of this information, and its semantic impact on the public and culture (Sui and Goodchild 2001).

The readers of the map, not having the possibility to participate in the elaboration of its language, can only imperfectly attempt to understand it. All the complexity, the creativity, the art in the broadest sense, is then concentrated on the side of the cartographer and his/her capacity to use a system of adapted graphic signs, following a logic centered on the effectiveness of communication.

These overly limiting approaches had already been noted, and various researchers had begun to make this model of cartographic communication more complex, developing new articulations and deepening its various elements. The discussions were numerous in the 1960s and 1970s, as can be seen, for example, in the international yearbooks of cartography[2].

The model proposed by Koláčný in 1969 (Figure 2.2) is one of the most detailed and cited. It represents a communication process that is also oriented, from map production to map use, but it suggests more complex information flows, with potential feedback loops. This model considers in particular the mental and cognitive context of the participants, both the cartographer and the user, but without pondering that it can be shared and linked to a partially common imagination. The model also integrates the transcription of geographical reality into visual signs, as well as the role of the map as a decision support tool to accompany the user's action.

In his model, Koláčný also considers the context of the map's operation (which will be developed later). According to this perspective, the author is motivated by the objective of transmitting information, which is selected and represented according to his or her knowledge and experience, with his or her means. Mirroring this, the readers will use the map with a certain goal of action, according to their knowledge and their experience of reading maps, all in a context of influential external conditions.

2 Trilingual magazine that accompanied the debates of the International Cartographic Association, between 1961 and 1985.

Among the new and important elements of this model is the presence of the notion of experience, on both sides of the model, which takes the training of the agents into account. It is completed by "psychological processes", which can imply, among other things, the more or less explicit will to refer to types of representations and styles, to visual aspects, which are recognizable and valued (traditional or, on the contrary, more original).

However, like the previous one, this model implies the general objective of transmitting "true" information drawn from reality, in an objective and revealing way. The challenge is also focused on the cartographer's capacity to interpret and represent the "real" world (according to the author, but we could use "observable" or "understandable") in a readable, efficient way, using a "cartographic language" shared by the largest number of people.

To complete these comments, we can call on Edney's (2019) recent work on the cartographic ideal. Indeed, the presuppositions underlying the communication model can be classified into what Edney calls "preconceptions", of which he offers an enlightening typology[3].

The preconceptions of individuality (the map is an individual act), materiality (a physical object) and, above all, efficiency and discipline/morality (a document that is meant to be efficient, elaborated by professionals who are supposed to be objective and neutral) are particularly appropriate in order to take a critical look at the model studied here.

From this vision of cartography as a communication tool, it is possible to deepen and enrich the analysis by completing this model with the integration of reflections developed later in semiotics, in information and communication sciences, and by taking into account the innovations, especially the technical ones, that have developed since its conception.

This is what we propose to do here, through the formalization by a series of graphs, making it possible to model the process of contemporary cartographic communication in a progressive way.

3 A French language review of the book offers a brief presentation (Jégou 2019).

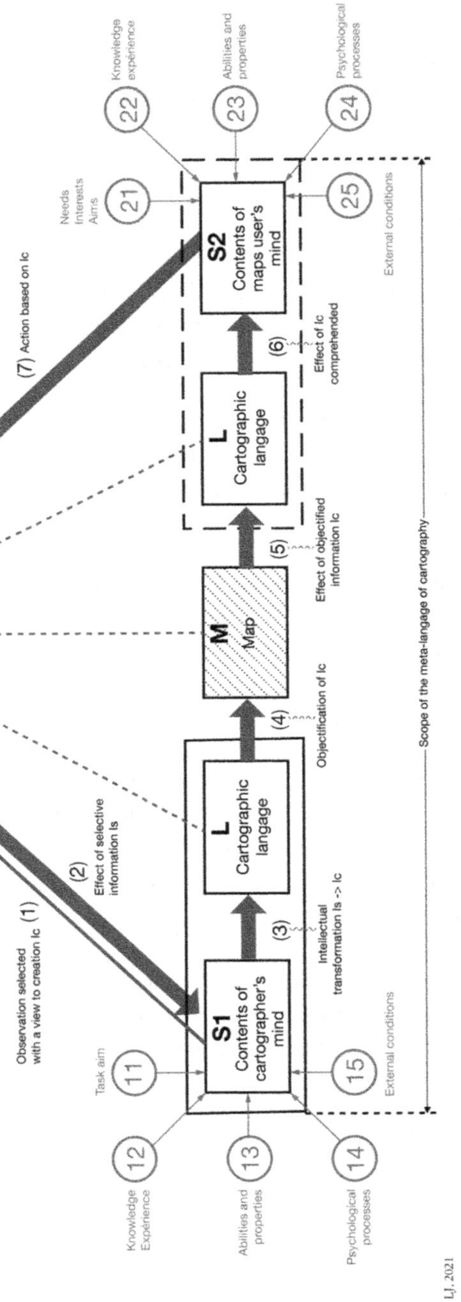

Figure 2.2. *Cartographic information (according to Koláčný 1969)*

2.3. Variations of the traditional model of cartographic communication

2.3.1. *The semiology of graphics, an essential element of the model*

In this first series of diagrams, the model of cartographic communication, shown as a starting point, is presented in Figure 2.3(a), as a simplification of Koláčný's version. This schematization makes it possible to propose progressive complexifications, starting from the simple situation of the production of a map as a document representing reality, by a specialist for an unspecified user, who is going to use it as a tool in his or her consideration of reality (navigation, understanding of a territory, etc.).

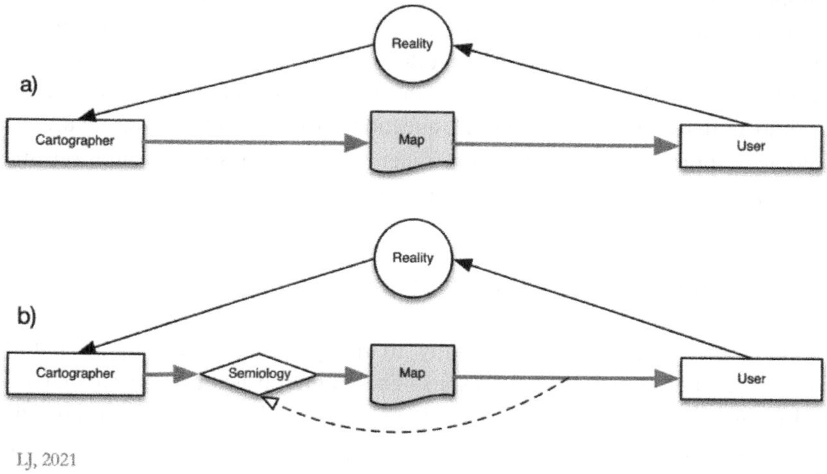

Figure 2.3. *Simplified version of the model and integration of the semiology of graphics*

We complete the diagram with the now traditional idea that the cartographer will use a "language" to encode his message, his perception of reality, in a map, which in this case is the semiology of graphics, as defined by Bertin (Figure 2.3b). This semiology is influenced and conditioned by the perception capacities (visual perception and interpretation) of the user (back arrow), as they have been analyzed and (potentially) taken into account by Bertin and his successors.

This model of cartographic communication has, over time, emerged as a paradigm in the teaching of thematic cartography, especially since its explicit validation in the work of Bertin (1978) and Tobelem-Zanin (2014), which is central in the field. Indeed, educators are encouraged by textbooks to ensure the proper understanding and use of the semiology of graphics by teaching how to read and implement it.

Most of the work in cartography is thus concentrated at the level of cartographic language, in other words, into the choice and representation of signs that will convey the message via the map object, given that its diffusion and reception cannot really be controlled. The semiology of graphics refers to a linguistic approach to cartography (Palsky and Robic 2000). Bertin particularly uses the expression of "visual grammar", in accordance with the dominant scientific theories of the time, such as structuralism and Saussurean linguistics. The semiology of graphics naturally constituted a major contribution to the methodology of thematic cartography. However, in addition to the limits already pointed out in the field of the psychology of perception (Green and Courtis 1966), the model of communication that it involves may have slowed down reflection on cartography by focusing on the notion of language, but also by making the research about the functioning of the map almost disappear from active questions of research. The return to a semiological approach, taking the more recent contributions of research on images into account, is beneficial, as we will try to show.

2.3.2. *Integrating the notion of intention of communication*

Limiting the function of cartography to the graphical encoding of reality is indeed very reductive for several reasons. This situation has been particularly well studied in linguistic semiotics[4] by Eco (1988) and Klinkenberg (2000) through the notion of intentionality and involuntary communication, its identification and the observation of the users' reactions.

Figure 2.4 shows this by breaking down the message of the "map" object into four parts:

– what the cartographer draws, encodes voluntarily (thus also, inversely, what is voluntarily not included);

– what he or she includes involuntarily, without realizing it, for example, through the reproduction of conventional forms (layout, fonts, colors, etc.) or what is forgotten;

4 In an effort to simplify the present text by not entering into the semantic debate, we will use semiology or semiotics according to the context of the sources, considering that they are equivalent for our general purpose.

– what the users perceive and identify in the map, even if they are mistaken, as well as invent, since an interpretation is produced;

– what they do not perceive in the map (whether it be intentional or not from its designer), due to lack of experience (especially with the cartographic language) or inattention, or else because of a mismatch with the purpose of the map.

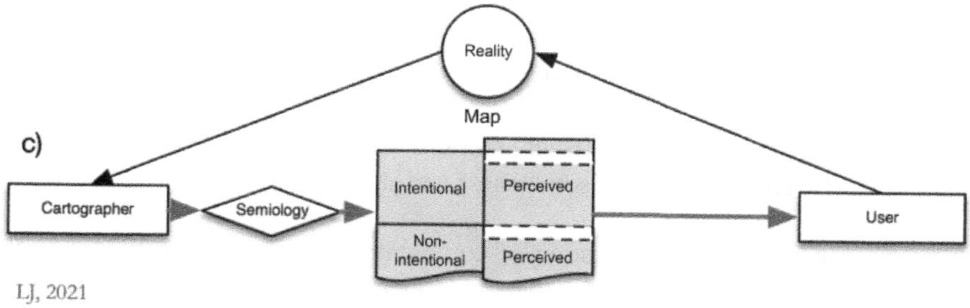

Figure 2.4. *Considering the intention of communication*

If we consider the example of a tourist map of a mountain valley, the cartographer will voluntarily include elements of reality such as the relief, tourist facilities, towns, villages and road infrastructures, in a pleasant style that invites people to come and enjoy these areas. Again voluntarily, the map producer may also decide not to draw certain parts of reality that are possibly unpleasant for tourists, but which nevertheless occupy a significant area in the landscape, such as factories or a quarry. Partly involuntarily, they may integrate conventional or stylistic elements, such as specific pictograms (signposts, symbols, etc.), color schemes, frames, etc., particularly in relation to the current trends in graphic design. They may also unintentionally forget certain touristic items due to a lack of information.

The users of this tourist map will perceive part of its content, perhaps not all of it, depending on their objective in consulting the map or their habits, as well as their experience with this type of map. They may also think they perceive elements that do not exist: for example, a passage between two rocky areas that is a blur in the image. Lastly, they may not understand a part of the map or make a mistake and misinterpret a pictogram or a color: Is the deep green fir tree the mark of an accessible, attractive place or rather a place that is impenetrable, to be discouraged? Is the green color related to the level of difficulty or the forested nature?

Beyond these situations linked to the contents, Eco is interested in the possibility that one can predictively encode signs which do not have to be perceived consciously by the user, but which can nevertheless influence interpretation. This perspective refers to a rhetorical or esthetic objective (in a desire to orient its public toward an emotion or to create an emotional reaction). Eco (1970) mentions in particular the example of comedy theater, the actor's performance, whose duplicity can be voluntarily exaggerated (to be detected as such).

In cartography, we can talk about maps that lie, be it voluntarily or involuntarily, to use the title of Monmonier's famous book (Monmonier 2019). The author may plan to deceive readers, or not realize that the map will often be misinterpreted. Authors can also integrate graphic elements that will eventually evoke certain ideas or feelings in readers, such as pleasant emotions associated with visual harmony or, on the contrary, an impression of seriousness and severity.

As we can see, the multiplication of communication situations considering the intentionality of both the cartographer and the reader makes the process more complex but, at the same time, allows us to propose a more detailed account of the forms of interaction between the parties involved in cartographic communication. This brings us closer to a more concrete consideration of communication, which takes the context and the reactions on the receiving end into account. This is the direction that several interesting research studies have followed, criticizing the overly simplistic, almost mechanistic vision of the traditional model.

2.3.3. *Context, operating modes and reception in cartography*

In the work of visual semiology that has continued since the 1970s, more pragmatic approaches have been developed. Here, we propose exploring the questions of context and operating mode, which complement each other and lead to the question of reception, that is to say, the way in which an image is perceived, interpreted and used by its users, to which the sciences of language and art have progressively become involved.

Continuing the famous work of Barthes (1964) on the rhetoric of images, the contribution of Joly (1994) on the semiology of images offers a summary and a relevant reading perspective to rethink the model of communication, in particular in the framework of cartography. The image is inevitably polysemous, it has multiple interpretations, but this is mainly because there are multiple spectators (Metz 1976). Moreover, reading an image is neither passive nor natural or instinctive, it requires study, especially for certain types of content. Reading an image is not immediate and comprehensive (Ware 2008; Saint-Martin 2011), and while the image is material, and therefore still, its context of interpretation can be changeable.

This notion of context is important in the shift toward a more pragmatic approach to the interpretation of images and particularly maps. With the same material support and the same participants in the act of communication, the meaning of a map can change completely depending on the context. We can take the traditional examples of decorative maps (placed on walls like paintings) or historical archives, in an educational situation (classroom) or official communication (background of a speech). In cartography, the idea that the context of use affects how a map is designed has become significant. Griffin et al. (2017) provide a useful summary.

In order to develop this notion, it is interesting to mention the recent work in general image semiotics around the theory of enunciation. By applying the concepts of linguistic semiotics to images, various authors have revealed the importance of focusing on the acts of production and observation of the images themselves, as well as on their context (Dondero 2020). In its production, the map object is defined, delineated, provided with the attributes that identify it and within a framework that is useful for its proper consideration. A certain context of use is therefore assumed.

For maps, these analyses thus invite us to pay attention to the contexts of production and use of maps, to social rules, as well as to "implicit contracts of communication". This brings us back to the notion of "cartographic modes" proposed by Edney (2019), which are ways of making, using and disseminating maps that are, in fact, quite varied. The same map can be used in very different contexts and for very different purposes: discovery, analysis or management of a territory, illustration, teaching of geography, propaganda, advertising, tourism and even decoration, as mentioned above. With the previously presented example of a tourist map of a mountain valley, several uses of this map are possible. It can be used to choose a holiday destination, as a road map or as a decorative vacation souvenir. This concerns both the purpose of the document and, more broadly, its context of use. This idea can be integrated into the model diagram in Figure 2.5.

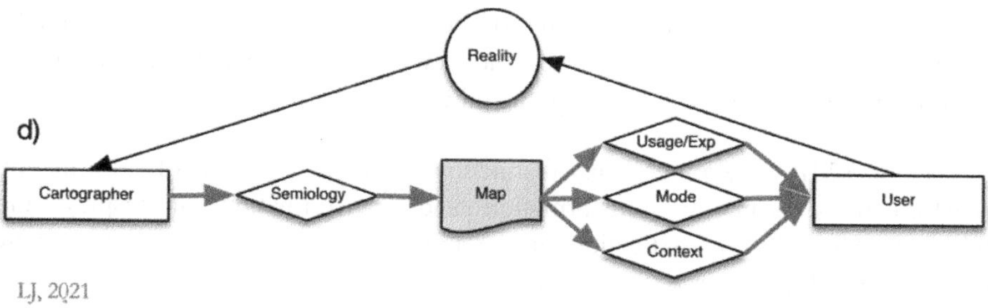

Figure 2.5. *Consideration of uses, modes and contexts in the model*

This way of understanding communication also allows for a better consideration of the reception (user) side of the model, and for the development of connections with what are called audience or reception theories (Littlejohn and Foss 2009). These theories arose from the need to examine the life of ideas interpreted by users, which were quickly realized to be neither stable nor completely predictable, as users are active in the interpretation, living in society and under different constraints, with varying backgrounds and experiences.

Ultimately, since the 1960s, all cartographic communication has been transformed by various technical innovations, starting with new means of map reproduction (color printing that has become accessible, fast, and accurate), then new media for diffusion (the screen, the web, cell phones, etc.), new forms of maps (dynamic, interactive, etc.) and new means of production (computing, geomatics). These technological changes have been described and commented on in several articles (Joliveau et al. 2013; Mericskay et al. 2018) and are questioned in greater depth in Chapter 5 and Chapter 6.

The techniques and modes of map production are evolving, and these changes influence the general modalities of their context of production (automation, connection with statistical databases, integration into observation systems, etc.). For example, new uses of maps, from the user perspective, have developed based on these innovations, such as interactive atlases or dashboards (MacEachren 1995; Sui and Goodchild 2011). The use of maps is becoming more diverse, commonplace, or democratized, in the sense that this use is no longer reserved for a small number of specifically trained individuals with access to complex, expensive tools that are only intended for a single practical, scientific, or educational use. Today, any person equipped with a cell phone potentially has a global geopositioning system by satellite, as well as access to interactive road maps, which adapt to their position and their uses (itineraries, searches for locations, services, etc.).

We can reflect on the integration of these new map functions in Figure 2.6 by adding intermediate phases on both sides of the map object: adaptation to new media by the cartographer on the one hand, and new needs and experiences linked to modernized equipment for the users on the other.

Cauvin (2016) mentions this distinction from the aspect of information transformations: semiotic transformation and transformation of display and diffusion. Also, the diversification of the forms of use of the map offers new perspectives, which deserve to be explored.

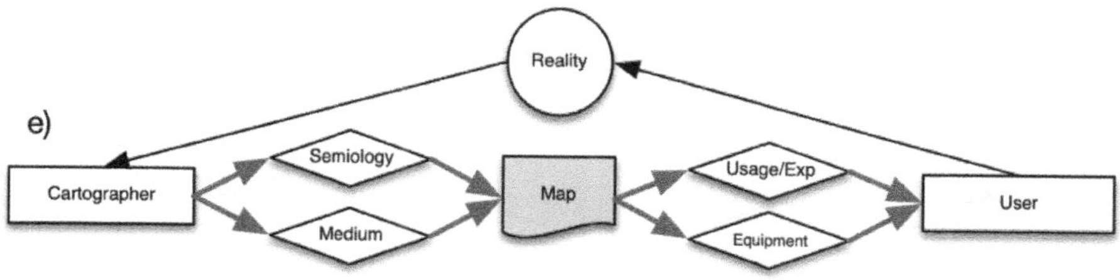

LJ, 2021

Figure 2.6. *Consideration of various media and uses in the model*

2.3.4. *Disruptions induced by the multiplicity of users*

The series of diagrams in Figure 2.7 aims to take into account the multiplicity of (new) forms of user involvement in maps, which are becoming more diverse.

Map readers, especially online, are no longer unique. Moreover, the contexts of production, the modalities of diffusion and the situations of map use are very diverse, and no longer refer to predetermined linear logics.

Thus, in the case of Figure 2.7(a), the right-hand side of the diagram is fragmented, the uses of the map are varied, particularly in their intensity, and the resulting actions in reality may be different. In *The Nature of Maps* (1976), Robinson and Petchenik already mentioned the variations in intensity of map "reception", going from a quick reading for a simple and specific action to a more thorough use to guide a more complex activity. Again, reference can be made to Griffin et al. (2017), for a reflection on the effect of considering varied uses and contexts on map design. The map no longer has a single undifferentiated use, centered around the utilitarian aspects of discovering, measuring or managing a territory, it now refers to multiple and protean interpretations and practices.

In Figure 2.7(b), we try to integrate the fact that, thanks to recent media, especially the web, which allow new forms of interaction between the user and the map, various types of uses go so far as to multiply the map. Cauvin (2016) mentions "a map object that has become multifaceted and scalable". For example, the map is no longer a paper document that is produced once and for all, it can take the form, on a screen, of a document equipped with an interactive interface, which users can explore (in which one can move around, change the scale), query (by searching for information by type, by theme, by text, etc.) and personalize (choice of the visualization scale, of the represented data, of the colors, etc.).

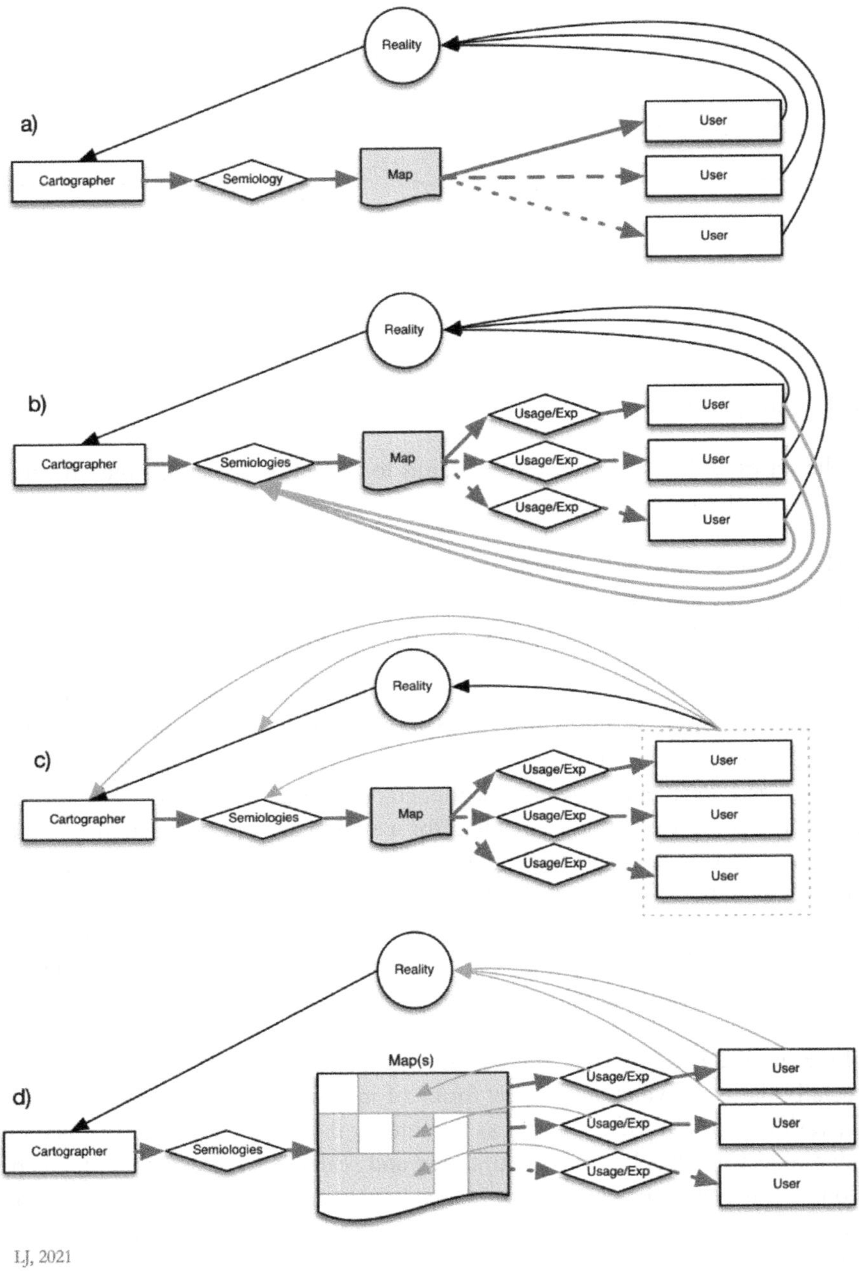

Figure 2.7. *Consideration of the involvement of different users*

These uses refer to different "user experiences", in other words, different ways of using the map, the latter being proposed with interactive functions that facilitate its modification (moving, zooming, filtering, list of base maps or layers and a time bar). These varied possibilities of use according to the cartographic products are quite significant, and they can therefore influence the choice of users, as well as weigh on the designers' preliminary reflection on the production and adaptation to the users' needs in the phase of representation of information, which was previously grouped under the term "semiology" (Figure 2.3). How the map works can therefore also be presented as a relational and dynamic process, shared between creator and user, composed of two-way information flows and feedbacks[5].

If we take this logic of audience intervention in map production further (Figure 2.7c), we can involve them in two additional steps. First of all, users can act at an earlier stage than before, namely in the phase of selecting the information to be represented, because as users, evaluators and purchasers of the map, they influence the cartographer's choices. In the case of cartographic applications on mobile phones and tablets, for example, which is a market that has become very competitive, users will choose the most functional map, but also the most accurate and appropriate from a wide range of products. The possibility of using a map on one's phone also implies that the map, both in terms of its formatting and content, is adapted to the place where one is, but also to our daily activities. The idea of egocartography put forward by Noucher (2017) is more widely discussed in Chapter 5.

The user can then, ultimately and to some extent, become a cartographer themself by participating in the creation and updating of represented data, like in open-source mapping projects (OpenStreetMap HotOSM[6]), naturalist inventories, or geographic crowdsourcing programs supervised by private companies or state bodies (Mericskay and Roche 2011). Beyond simple contribution, map users are also able to train themselves in cartography using computer tools that have become more affordable (free and open-source), accessible, and similar to those of professionals. This new form of relationship between the user and the map could also account for the fact that the cartographer are often, first and foremost, their own users, with geographers often making their own maps for exploratory purposes, to discover new information. Lastly in this category, we find the projects of map "co-construction" between cartographers and users, which are experiences that have

5 See Chapter 7, which deals extensively with these questions.

6 A project for the participatory production of geographic data (based on OpenStreetMap) in the context of emergency situations (natural disaster, health crisis, political crisis, etc.).

notably developed in the course of critical cartography, such as in the support of urban projects that integrate the participation of the inhabitants[7].

The last variation (Figure 2.7d) considers the possibility offered by the personalization techniques of the map by the user (on web media) via the choice of the mapped elements and the aspect of their representation among a list of possibilities provided by the cartographer (who has then become an applications developer).

Although the choices of representation modes are often preconfigured in order to take the good practices of the "cartographic language" into account, the final representations themselves and their combinations, the base maps, the scales and the graphic styles can be left to the users to choose.

The map document is no longer fixed in advance, nor is its content fixed once and for all. This brings us back to the previously mentioned idea of a "patchwork map", from the standpoint of intentionality and interpretation, or the idea of the map as a message reassembled by the users.

Examples of this type of cartographic co-construction can be found in online tools such as UMAP from OpenStreetMap, MapMe and ZeeMaps, where several users can collaboratively modify the style of the maps, add data or comment on the content. Institutional geoportals, such as those of the IGN or the Swiss Federal Office of Topography, as well as private platforms such as Google Maps or Microsoft Bing Maps, allow users to build their own maps in just a few clicks (base map, displayed information, graphic styles of displayed elements, etc.).

The search for the inclusion of different, or even innovative, forms of cartographic communication in the model thus allows the latter to evolve and to attempt describing these new functions. It is also an indication of the interest in exploring the question of implicit or involuntary communication, which must play a significant role in the other operating mode of the map: the esthetic domain.

2.4. How to integrate the esthetical map function?

It is a widely shared observation: the map, as a graphic object, also has an esthetic function, that is to say, it appeals to the senses and to a qualitative,

7 Palsky (2018) provides a synopsis. Noucher (2017) develops the issues, as well as describes and analyzes several examples. The critical approach in cartography is also well represented in the *Atlas critique de la Guyane* (Polidori and Noucher 2020).

emotional appreciation. We quote here an extract from the definition of the map, given by Zanin for the online encyclopedia Hypergeo:

> Cartography is an art, because as a mode of graphical expression, the map must present qualities of form (aesthetic and didactic, thanks to the clarity of the line, its expressiveness and its readability) in order to exploit the visual capacities of the reader. This requires choices on behalf of the designer and the producer. (Zanin 2021)

Bertin (1978) even considers it to be an art. Keates (1996) reinforces this by stating that a good map is an effective combination of expressive artistic design and spatial data to inform, but to also affect the reader.

The map being by definition and vocation an expressive, communicative representation, its primary role is to be interpreted by its users to evoke ideas about a territory and its spatial phenomena. The map is a particular object, possessing a power of expression whose multiplicity and depth of possible interpretations refer to both esthetic and emotional questions (Griffin and Mcquoid 2012; Muehlenhaus 2014).

According to Eco (1970), the semiotic function also relies on the activation of the connoted meanings in the reader. An image can provoke secondary interpretations that are emotional, uncontrollable or even instinctive. The Groupe μ has subsequently developed an in-depth research on the semiotics of visual signs (1992), which can be adapted to the map object.

In order to study the esthetic aspect in cartographic communication, the art sciences, guided by semiotics, constitute a rich perspective (Jégou 2016). It is perhaps also one of the fields in which cartographic innovation could be quite visible, through the proposal or consideration of new forms, new and original visual aspects, expressing creativity.

Its comparison to poetry can then be considered. It is perhaps an interesting way to study the abundant production of "neo-cartographers" (Palsky 2013) and Geoweb cartographers (see Chapter 6). In order to remain within our approach to image semiology, we will only focus here on visual aspects, even if the esthetic aspect of maps can take other forms that are well explored by the current course of sensitive mapping (emotions in situ, sound and smell environment, interactions with the place and its memories, etc.).

Starting again from the original communication model, we now seek to progressively integrate this esthetic aspect into the models that are already identified.

2.4.1. *Consideration of esthetics from the cartographer's side*

The first diagram (Figure 2.8a) considers the dual character of the map (tool/artwork) from the cartographer's point of view, by complementing their capacity to encode reality through graphic language with an ability to produce an object with esthetic properties, in other words, to act as an artist. Even if the definitions of this function are far from being unanimous and stable, we will choose Pouivet's (2010) definition for its practicality, which retains the intentional character of the production of the work of art, as well as its recognition as being artistic by the public. According to this first approach, cartographers use both their knowledge of semiology and art to produce a document that graphically expresses reality in two complementary ways. One is explicit based on a formalized code, and the other is partly implicit, which functions through evocation, expression and the suggestion of esthetic properties. It should be noted that these two modes are characterized by the idea of metaphors, which is central in the field of semiotics, with extensions to the cognitive sciences (Lakoff and Johnson 1980).

This raises the question of the symmetry of the diagram, for example through the integration of modifications and extensions of the model around work on graphic representation (Suwa et al. 2001). If the cartographer draws elements from reality in order to represent and transmit them, what happens on the artistic side of the cartographic functioning? The seemingly logical choice aims to complement the reference to reality with a call to the imaginary and imagination, which would constitute the source of the evocations and emotions that would be expressed by the artistic side of the maps (Figure 2.8b).

Of course, imagination can touch upon reality and be dependent on it, but always maintaining a distance, a metaphorical representation. The map will trigger the production of mental images about the represented territory: the material functioning of the map, the representation of reality via a conventional language, is completed by an act of creating-expressing mental images that are more or less guided by free association. The reader will be stimulated, partly unconsciously, in their thinking and action, which is one of the ways the map works (MacEachren 1995).

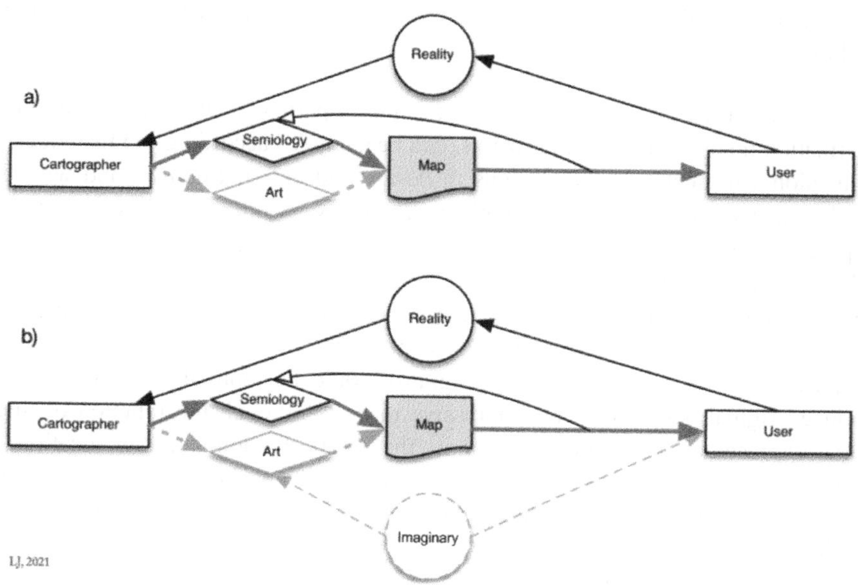

Figure 2.8. *Consideration of the esthetic functioning of the map*

With the example of the tourist map of a mountain valley, the evocation of places and their representation (signs, styles) will potentially provoke mental images of the valley, its landscapes and the tourist activities that one anticipates or remembers. A vacation souvenir map makes full use of this capacity. The great outdoors, the landscapes, will be evoked explicitly by the representations of the landforms and the winter sports equipments, but also, more metaphorically, by the colors, the dominant white of the snow, the presence of pictograms, the images of the forest, of characters enjoying winter sports, etc.

The expressive capacity of color is also a particularly interesting element to develop. Although the map often mobilizes colors gradients to graphically transcribe a relative variable, these same colors recall other images and colored sensations, for example of elements of reality associated with these colors by memory, by the very functioning of our cognitive representations by resemblance. In addition to the information explicitly decoded through the comprehension of graphic signs, imagination makes emotions emerge by reacting to various characteristics of the image which one seeks to interpret. Color is one of the most powerful signs for this dual expression, as shown by Christophe (2009) or Fabrikant et al. (2012), in their work on the use of color palettes from pictorial works in cartography. This esthetic functioning has been described in visual semiotics by the Groupe μ (1992) as an empowerment of the *plastic* sign, functioning by itself and no longer as a simple attribute of the signifier.

The inventiveness of the “new” maps, found in particular in the press, in online information media or in social networks, can be analyzed according to this model, in connection with technical innovations in design and data-journalism, and with the development of a sensitivity to graphical creativity. Numerous atlases, computer graphics and illustrations, especially on the web, demonstrate a desire to develop the esthetics of representation and formal creativity[8]. It would be interesting to question the motivations of this interest for the visual aspect which, in a competitive, commercial (press) or social interaction context based on more and more images (social networks like Twitter and Instagram), could be motivated by the need to produce original, surprising and impressive maps[9]. Although the function of these spatial representations is often far from the scientific domain, the vocation of information and educational intent are quite often present. We can establish a link here with the idea that the cartographic process is also a “performance”, in the artistic sense of the term (Perkins 2003).

2.4.2. *Consideration of esthetics from the user’s side*

The diagram in Figure 2.9 incorporates the downstream part of the operation, on the users’ side, where the objective of using the map and the experience of this exercise will be achieved through sensibility to the esthetic aspects.

This stimulation of the reader’s imagination is only potential, because it depends on the subject, even if the appreciation of certain esthetic forms is more or less stabilized or set in styles and fashion trends (ranges of colors, types of lines and hatching, fonts with or without a “serious” appearance, etc.). Mountain tourist maps, for example, have a characteristic “pictorial” style, and often propose perspective views (instead of top views, which are more traditional in cartography).

The cartographer can try to consider the particular sensibilities of the users in the way the artistic side of his or her work will be designed, but there is no guarantee of success. The development of the use of cartographic forms by artists helps to validate this esthetic functioning (Harmon and Clemans 2009). Ultimately, the relationship to imagination can have a dual meaning, for both the cartographer and the users: the map, as an original composition, can also influence their imaginations.

8 See the selection of books proposed by the *Mappemonde* journal for the past few years or more experimental books such as *Terra Forma* by Grégoire et al. (2019).

9 See Chapter 3, which also addresses this matter.

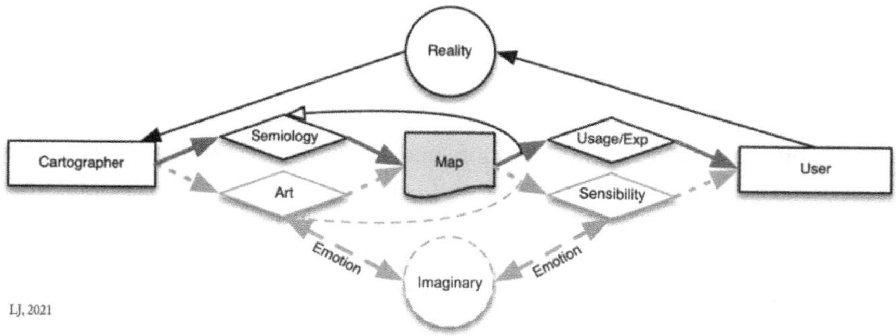

Figure 2.9. *Esthetic operation, downstream development*

The esthetic dimension of the map can also go beyond the visual aspect, to engage the relationship more dynamically with the users, in a more or less formalized "game" that will encourage a more active and playful relationship with the document. This notion of a game was explicitly expressed by the authors of *We Are Here: An Atlas of Aotearoa* (Denee and McDowall 2019), an atypical atlas of New Zealand. They claim that they want to produce unusual, surprising images, in order to encourage the reader to look for geographical landmarks, omitting some common map "habits" (no base map, but the content, by its very form, gives clues to the geography), or to promote surprising metaphors between the territory and the theme represented (such as that of the anatomy of a circulatory network by a transit map) by showing an initially ambiguous image that does not even look like a map. Their objective is, first of all, to overcome the obstruction (or boredom) that can be caused by a map that is too complex, overloaded with information and too traditional in appearance.

The consideration of multiple users, who can intervene in the map design and its concrete configuration to adapt it to their needs, as mentioned previously (Figure 2.7d), could naturally be mentioned here, taking esthetic aspects into account. The corresponding schema would, however, be very complex. As with the explicit signs of the maps, the configuration of the implicit, plastic signs, of the meanings expressed by the esthetic properties, can be anticipated from its conception as changeable and customizable by the users of the maps. This is the particular research question engaged by Christophe (2009) and Ory (2016) about the personalization of topographic map styles. In the same way as in art history, one can imagine a categorization of maps by their visual styles: maps that are rather neutral, uncluttered and readable for scientists, the military and institutions, maps that are humanized and made more sensitive for critical or militant cartographers, for example with the use of hand-drawn pencil sketches (Rekacewicz and Tratnjek

2016), maps which are expressive and moving for artists, etc. The research perspectives mobilizing art history tools on maps are numerous and fruitful, as already shown in the work of Carani (1995).

As mentioned above, taking emotion into account in cartography also echoes current trends, like radical cartography (against dominant positivist and liberal ideas, linked to critical geography) and sensitive cartography. For the former, the stimulation of the reader, as mentioned by MacEachren, is then contemplated through information on social issues and as an incentive to take concrete action. Rekacewicz discusses the idea of mapping with emotion in an illuminating interview (Rekacewicz and Tratnjek 2016). We can also directly connect this idea of the map as a medium of emotion and transmission of the imagination, with the work of sensitive mapping (Mekdjian and Olmedo 2016).

2.5. Which places of innovations are in the cartographic communication model?

To conclude this series of diagrams aimed at questioning, breaking down, enriching and modernizing the "traditional" model of communication in cartography, we can attempt to summarize the parts of this model that are conducive to change, referred to as "places of innovation" (Figure 2.10).

2.5.1. *The designing of the map*

The first place of innovation implied by information theory and the communication model, at a time when the physical medium of the map was not questioned, was in the processing, encoding and representation of reality in the map by the cartographer. For a long time, this "period of time" of the model concentrated on research efforts (in visual perception, graphic design, semiotics, geomatics, etc.) to formalize the methodology, mainly in France around Bertin's semiology of graphics, but also around Robinson and Petchenik in the United States. Computerization, the development of GIS and more recently the multiplication of connected objects and the emergence of research on big data have largely multiplied the possibilities in this "upstream" part of the model and the capacities of production, management of geographic data and representation of reality, all while encompassing previously developed methods (Robinson et al. 2017).

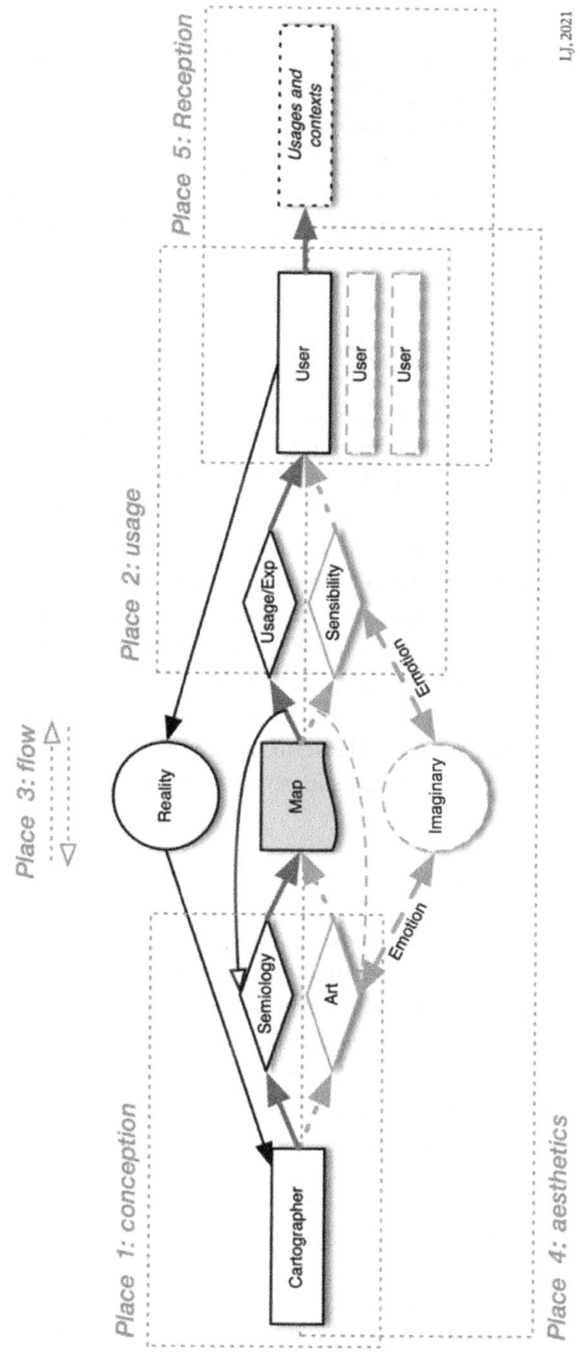

Figure 2.10. *Places of innovation within a renewed model. For a color version of this figure, see www.iste.co.uk/mericskay/cartographic.zip*

2.5.2. *Using the maps*

The second place of innovation is further downstream, at the time of the consultation and application of the map by its users, as the medium of the map changes with technical progress and its uses are modified. The possibility of using color more freely, then screen support, interactivity and lastly, mobile devices allowing the interaction of a large number of users, all fall into this category. These new ways of using maps have, in turn, greatly influenced the production and design processes of maps.

2.5.3. *Diversification of communication flows*

The third category of innovation does not involve a specific point of the model, but its general organization and the information flows that cross it. The users of the map are no longer simply undifferentiated readers who use it in a generic way, but are now multiplying and promoting various uses, that are not always anticipated, which will in return influence the design and production of the maps. The users can become citizen-sensors or data providers to feed the maps or even cartographers themselves. The map is thus no longer a fixed object on a fixed medium, but a dynamic and evolving production based on a collective construction and enrichment approach. In the case of sensitive cartography, the map is no longer a univocal authoritative document that shows reality, but the co-constructed expression of an experience, a media in the full sense of the term, a combined support, agent and resource, result and process, and even an artistic performance tool.

2.5.4. *A new esthetic functioning*

Considering an esthetic functioning provides the material for a fourth type of potential innovation in cartography, essentially influencing the form, the visual aspect of the map, in the direction of a greater qualitative appreciation. This perspective also allows for a convergence with media, materials where creativity and the expression of the imagination can be developed more widely (works of art, literature, painting, graphics, etc.). The map then loses its seriousness, its supposed objectivity, but gains a wider audience.

2.5.5. *Focus on the reception of the maps*

The link with artistic disciplines in this reflection on models of functioning leads us to propose one last line of research: the reception of maps. Within the works that aim at understanding and improving cartographic efficiency (in the sense of communication), many efforts have focused on the stages of encoding and visual perception.

Thus, we could, for example, try to better understand how the map acts as a media object, in its uses and its reception by its public. Jacob (1992) clearly points this out in his reference work, where he emphasizes that a map "is perhaps defined less by formal features than by the particular conditions of its production and reception, by its status as an artifact and mediation in a process of social communication". Servais (2012) proposes some avenues in information science. In particular, he puts forward the idea that the map can be a "media text" that has its own internal constraints (its content can direct the reading through standardized uses and expectations), such as issues of document type and socioeconomic context (the dominant social code will guide the reading of the text without the intervention of the text itself, for example, the role of the map as a power relay for the control of a territory).

These analyses echo ideas that have already been explored in cartography, notably by Harley (1989), but perhaps from a too directly utilitarian angle. The proximity with the communication studies appears to be very beneficial in terms of new ideas on cartographic functioning, as these disciplines have noted. The issue of the *Communication & langages* journal, entitled "The map, a medium between semiotics and politics" (Robert and Souchier 2008), is a good example of the theoretical and methodological contribution in considering cartographic communication in a novel way.

2.6. Limitations and conclusion

While the communication model allows us to describe the essential functions of the map, it also restricts it to the role of an instrument for transmitting information, with the aim of taking action. This function is important, but it does not take into account the development of information and communication technologies, computer science, the web and the new uses of the map that are linked to it, going beyond the sole purpose of communication and the relationship between design and use.

Moreover, this approach is restricted to representational cartography that produces documents. With the development of spatial databases, the objective of representation and transmission of information is now added ex-post to that of data storage, analysis and management. The data are rarely used to produce maps, and instead used to feed indicator tables, spatial analyses and itineraries (Joliveau 2011). The development of new means of acquiring, structuring and processing data has concentrated innovation in the field of geographic data, with the development of GIS. Their structured distribution is carried out via spatial data infrastructures on the Internet, whose representations are partly automated (so the cartographer no longer designs the image, no longer has detailed control over the final aspect of the result[10]). Reflections on cartographic communication still take little account of these fields of activity, which are nevertheless directly linked to current cartography, and one may ask how to better link reflections on cartography to the prolific issues surrounding spatial data and their processing[11].

In addition, we can mention two general cases in which the transmission of information to guide a concrete action takes a back seat. The first is where the primary objective of the map is esthetic: representing non-existent, imaginary places, wanting above all to make the user feel emotions, to create a sensory universe. We think of the example of artistic works that take up certain cartographic forms or intentions. The book by Harmon and Clemans (2009) constitutes a good overview of this type of work. We also think of the works presented in the context of the "Time, Art & Cartography" conference or the "Cartologies" exhibition in Strasbourg in 2016. The second case is when the aim of the activity is a showcase to demonstrate a certain ability to produce a complex document, some kind of feat of synthesis and design. One could typically find this kind of production in some national atlases. The *Ukraine in Maps* atlas is a particularly enlightening example. However, more recently, certain geovisualizations are more of a technical "smoke and mirrors" with regard to this objective than the effective transmission of geographical knowledge, concerned with the achievements in this field (Bahoken and Jégou 2015).

To conclude, we can say that the communication model, which is central in the cartographic thought process, can be used as a starting point, to be criticized, modified, extended and declined with regard to the evolution of techniques and uses in maps. The analyses proposed here, through the formalization of more or less complex models, do not claim to cover the evolution of cartographic functioning exhaustively, especially in its recent multiplications in data processing, but can, however, constitute a way of offering a step back on different current trends and research methods. Innovation, in the prism of this model and its transformations, can be taken into account and encouraged in several ways: traditionally, by improving the representation of reality in an understandable and faithful document, as well as by adapting it to technologies that are themselves innovative, generating new uses

10 Conceptual and technical standards for cartographic representation are being discussed at the Open Geospatial Consortium. Available at: www.lejournal.cnrs.fr/articles/nouvelle-donne-pour-les-cartes.

11 See Chapter 3.

and therefore becoming more attractive. However, it is also important to note the role of innovation in stimulating map users. Dynamic and interactive online maps and geovisualization applications stimulate the imagination, as well as the actions of users, encouraging them to better understand and interact with the representations, and making them participate in their design or configuration, in order to personalize them[12]. Ultimately, innovation offers researchers new ways of considering and analyzing the functioning of the map.

2.7. References

Bahoken, F. and Jégou, L. (2015). Visualiser 15,3 trillions de dollars US d'exportations commerciales mondiales, and what else? *Mappemonde*, 117 [Online]. Available at: http://mappemonde-archive.mgm.fr/num45/internet/int15101.html.

Barthes, R. (1964). Rhétorique de l'image. *Communications*, 4(1), 40–51.

Bertin, J. (1967). *Sémiologie graphique : Les diagrammes – Les réseaux – Les cartes*. Mouton-Gauthier-Villars, Paris.

Bertin, J. (1978). Theory of communication and theory of the graphic. *International Yearbook of Cartography*, 18, 118–126.

Carani, M. (1995). De l'histoire de l'art à la sémiotique visuelle. *Style et Sémiosis*, 23(2), 37–49.

Cauvin, C. (2016). La sémiotique en cartographie au XXIe siècle ? *Cartes & géomatique*, 229/230,17–31.

Christophe, S. (2009). Aide à la conception de légendes personnalisées et originales : proposition d'une méthode coopérative pour le choix des couleurs. PhD Thesis, University of Paris-East, Marne-La-Vallée.

Collectif (2005). Cartography and power: The "post-structuralist turn", 1982–1991. *Cartographica*, 40(1/2), 85–111.

Denee, T. and McDowall, C. (2019). *We Are Here*. Massey University, Wellington.

Dondero, M.G. (2020). *The Language of Images: The Forms and the Forces*. Springer, New York.

Éco, U. (1970). Sémiologie des messages visuels. *Communications*, 15(1), 11–51.

Éco, U. (1988). *Le signe*. Labor, Brussels.

Edney, M.H. (2019). *Cartography: The Ideal and Its History*. University of Chicago Press, Chicago.

12 For more on this matter, see Chapters 6 and 7.

Fabrikant, S.I., Christophe, S., Papastefanou, G., Maggi, S. (2012). Emotional response to map design aesthetics. In *7th International Conference on Geographical Information Science*, Columbus.

Green, R.T. and Courtis, M.C. (1966). Information theory and figure perception: The metaphor that failed. *Acta Psychologica*, 25, 12–36.

Grégoire, A., Arènes, A., Aït-Touati, F. (2019). *Terra Forma, manuel de cartographies potentielles*. B42, Paris.

Griffin, A. and Mcquoid, J. (2012). At the intersection of maps and emotion: The challenge of spatially representing experience. *Kartographische Nachrichte*, 62(6), 291–299.

Griffin, A., White, T., Fish, C., Tomio, B., Huang, H., Sluter, C.R., Meza Bravo, J.V., Fabrikant, S.I., Bleisch, S., Yamada, M., Picanço, P. (2017). Designing across map use contexts: A research agenda. *International Journal of Cartography*, 3(1), 90–114.

Groupe µ (1992). *Traité du signe visuel*. Le Seuil, Paris.

Harley, J.B. (1989). Deconstructing the map. *Cartographica*, 26(2), 1–20.

Harmon, K. and Clemans, G. (2009). *Map As Art: Contemporary Artists Explore Cartography*. Princeton Architectural Press, Hudson.

Jacob, C. (1992). *L'empire des cartes : approche théorique de la cartographie à travers l'histoire*. Albin Michel, Paris.

Jégou, L. (2016). L'imagination esthétique dans la conception graphique des cartes : proposition de typologie illustrée. *Cartes et Géomatique*, 229/230, 231–248.

Jégou, L. (2019). Cartography: The ideal and its history. *Mappemonde*, 127 [Online]. Available at: https://journals.openedition.org/mappemonde/1711.

Joliveau, T. (2011). Le géoweb, un nouveau défi pour les bases de données géographiques. *L'Espace géographique*, 40(2), 154–163.

Joliveau, T., Noucher, M., Roche, S. (2013). La cartographie 2.0, vers une approche critique d'un nouveau régime cartographique. *L'Information géographique*, 77(4), 29–46.

Joly, M. (1994). *L'image et les signes. Approche sémiologique de l'image fixe*. Nathan, Paris.

Keates, J.S. (1996). *Understanding Maps*. Addison Wesley Publishing Company, Glasgow.

Klinkenberg, J.M. (2000). *Précis de sémiotique générale*. Le Seuil, Paris.

Koláčný, A. (1969). Cartographic information, a fundamental concept and term in modern cartography. *The Cartographic Journal*, 6(1), 47–49.

Lakoff, G. and Johnson, M. (1980). *Metaphors We Live By*. Chicago University Press, Chicago.

Littlejohn, S.W. and Foss, K.A. (2009). *Encyclopedia of Communication Theory*. SAGE, Los Angeles.

MacEachren, A.M. (1995). *How Maps Work: Representation, Visualization and Design*. The Guilford Press, New York.

Mekdjian, S. and Olmedo, E. (2016). Médier les récits de vie. Expérimentations de cartographies narratives et sensibles. *Mappemonde*, 118 [Online]. Available at: http://mappemonde.mgm.fr/118as2/.

Mericskay, B. and Roche, S. (2011). Cartographie 2.0 : le grand public, producteur de contenus et de savoirs géographiques avec le web 2.0. *Cybergeo: European Journal of Geography* [Online]. Available at: https://journals.openedition.org/cybergeo/24710.

Mericskay, B., Noucher, M., Feyt, G. (2018). Usages collaboratifs de l'information géographique. *Revue internationale de géomatique*, 28(1), 7–14.

Metz, C. (1976). L'étude sémiologique du langage cinématographique : à quelle distance en sommes-nous d'une possibilité réelle de formalisation ? *Communications*, 2(2), 187–200.

Moles, A.A. (1964). Théorie de l'information et message cartographique. *Sciences, Revue française des sciences et des techniques*, 5(2), 11–16.

Monmonier, M. (2019). *Comment faire mentir les cartes*. Autrement, Paris.

Muehlenhaus, I. (2014). Going viral: The look of online persuasive maps. *Cartographica: The International Journal for Geographic Information and Geovisualization*, 49(1), 18–34.

Noucher, M. (2017). *Les petites cartes du web : approche critique des nouvelles fabriques cartographiques*. Éditions Rue d'Ulm, Paris.

Ory, J. (2016). Connaissances pour la conception et la perception de styles topographiques. PhD Thesis, University of Paris-East, Marne-la-Vallée.

Palsky, G. (2013). Cartographie participative, cartographie indisciplinée. *L'Information géographique*, 77(4), 10–25.

Palsky, G. (2018). Construire des cartes en commun. La participation, nouveau paradigme de la cartographie. In *La carte invente le monde*, Picouet, P. (ed.). Presses universitaires du Septentrion, Lille.

Palsky, G. and Robic, M.C. (2000). Aux sources de la sémiologie graphique. *Cybergeo: European Journal of Geography* [Online]. Available at: https://journals.openedition.org/cybergeo/554.

Plantin, J.C. (2014). L'avènement de la carte comme médiation. *Questions de communication*, 25(1), 309–326.

Polidori, L. and Noucher, M. (2020). *Atlas critique de la Guyane*. CNRS, Paris.

Perkins, C. (2003). Cartography: Mapping theory. *Progress in Human Geography*, 27(3), 341–351.

Pouivet, R. (2010). *L'ontologie de l'œuvre d'art*. Vrin, Paris.

Rekacewicz, P. and Tratnjek, B. (2016). Cartographier les émotions. *Carnets de géographes*, 9 [Online]. Available at: http://journals.openedition.org/cdg/687.

Robert, P. and Souchier, E. (eds) (2008). La carte, un média entre sémiotique et politique. *Communication & Langages*, 158.

Robinson, A.H. (1960). *Elements of Cartography*. Wiley, New York.

Robinson, A.H. and Petchenik, B.B. (1975). The map as a communication system. *The Cartographic Journal*, 12(1), 7–15.

Robinson, A.H. and Petchenik, B.B. (1976). *The Nature of Maps: Essays toward Understanding Maps and Mapping*. The University of Chicago Press, Chicago.

Robinson, A.H., Demšar, U., Moore, A.B., Buckley, A., Jiang, B., Field, K., Kraak, M.J., Camboim, S.P., Sluter, C.R. (2017). Geospatial big data and cartography: Research challenges and opportunities for making maps that matter. *International Journal of Cartography*, 3(1), 32–60.

Saint-Martin, F. (2011). *Sémiologie du langage visuel*. PUQ, Quebec.

Servais, C. (2012). Les théories de la réception en SIC. *Cahiers de la Sfsic*, 8(1).

Shannon, C.E. and Weaver, W. (1949). *The Mathematical Theory of Information*. University of Illinois Press, Urbana.

Sui, D.Z. and Goodchild, M.F. (2001). GIS as media. *International Journal of Geographical Information Science*, 15(5), 387–390.

Sui, D.Z. and Goodchild, M.F. (2011). The convergence of GIS and social media: Challenges for GIScience. *International Journal of Geographical Information Science*, 25(11), 1737–1748.

Suwa, M., Tversky, B., Gero, J.S., Purcell, A.T. (2001). Seeing into sketches: Regrouping parts encourages new interpretations. In *Visual and Spatial Reasoning in Design II*, Tversky, B., Gero, J.S., Purcell, T. (eds). University of Sydney, Sydney.

Tobelem-Zanin, C. (2014). Cartographie et analyse territoriale. HDR Thesis, Université Paris-Diderot, Paris.

Ware, C. (2008). *Visual Thinking for Design*. Morgan Kaufmann, San Francisco.

Zanin, C. (2012). Cartographies. *Encyclopédie en libre-accès Hypergeo* [Online]. Available at: https://www.hypergeo.eu/spip.php?article371.

3

Communication Challenges in Reproducible Multi-representation

Nicolas LAMBERT, Timothée GIRAUD and Ronan YSEBAERT
RIATE, CNRS, University of Paris-Diderot, France

3.1. Introduction

All cartographic work is, above all, a matter of choice. This well-known adage by cartographers may seem trivial, but it deserves to be explained and discussed when the point of interest is the cartographic communication process. Indeed, at a time when geographic data are more numerous and diversified, and cartographic tools are more accessible and increasingly sophisticated, it is important to dwell on this question of choice in map construction. The objective of this chapter is to demonstrate, through exploration and practice, that each map is the result of decisions made by the cartographer: the selection of information, the graphical choices, the level of methodological complexity; taking these into consideration makes it possible to assert a communication strategy in order to effectively transmit a message to a given audience.

In this chapter, through examples, we explain that maps are as much scientific objects as they are formidable weapons of communication, because of the techniques and theoretical basis on which they are constructed. To support this demonstration, we use the same conventional geographical data, a population grid on a global scale, to show that cartographic communication can be expressed in multiple ways and return various forms of representations. We wish to prove that there is no mechanical relationship between the data and its graphic expression, and that every map is the result of an intention. In reality, it is this cartographic intention that governs all the decisions made during the design of the map, more than the technical and methodological aspects, ultimately resulting in an intelligible, organized and communicable document.

To do this, we have decided to anchor our demonstration in a traceable, transparent and reproducible process. Each map is thus associated with the R source codes with which it was generated.

The first part will deal with the general challenges of cartographic communication. After a short introduction, we will address both the scientific and political aspects of maps. Particular attention will be paid to the rhetorical aspects mobilized in thematic maps, which make their traceability all the more necessary.

Then, in a second part, we will propose the practical application of these concepts. From a single set of data on the world's population, a series of 13 maps will be presented and discussed, each one being created according to a specific intention and methodology, and conveying its own message. Finally, we will gather a certain number of graphic processes that echo the treated theme, with the aim of increasing the rhetorical power of the maps.

3.2. Searching and communicating with maps

3.2.1. *The map as a product of a scientific process*

Cartography is a scientific field that rests on a methodological base defined by theories and practices that continues to elaborate a "mode of use" (Brunet 1987) of its purpose between technique and visual mediation (Ferland 2000). It allows us to fix the invisible and prompts us to think about what surrounds us by extracting us from the individual schemes of perception. In this sense, it makes space tangible. Much more than a simple assembly of statistical figures in a table, the map makes it possible to bring out and visualize structures, logics and spatial dynamics, which are keys to explaining and understanding the world. We think here of Cooper and Snow's maps which, in 1858, made it possible to establish the link between the cholera epidemic that was raging in London and water contamination (Cameron and Jones 1983; Koch and Denike 2009).

Through this example and many others, we can see that the map (and by extension geographic information systems and online cartography) is the product of a rational approach that facilitates access to knowledge by offering a spatial reading.

For the social scientist that has the ability to format and model, the map is a tool that is widely used to explore data, test hypotheses and confront approaches with peers.

In this sense, the use of cartography contributes to scientific construction. This tool is also particularly adapted to synthesize and communicate knowledge and results to non-expert audiences, making the latter capable of reading and understanding these graphic documents.

3.2.2. *A highly political tool*

This formalization of space that is the map is rarely free or unbiased (Lacoste 1976). As Bahoken (2020) points out, the mapped motif is not silent; it is endowed with a meaning that reflects the intention of the designer (or author) of a map. The representation of migration is a good example to illustrate this idea (Figure 3.1).

In the hands of the Frontex agency[1], the map is used to determine where the EU's external borders are most exposed to the influx of migrants and, consequently, where they should be "protected" through a deployment of troops. This vision is also often affirmed by strong cartographic choices: colors, arrows and words (Houtum and Bueno Lacy 2020).

For NGOs, maps instead serve as a way to denounce the absurdity and inefficiency of migration policies, as well as practices that are contrary to the most basic human rights. In the hands of nationalist political leaders, they can be used to plan and monitor the construction of walls on their borders.

For activists, maps take on a tactical aspect used to dig tunnels or determine discrete passages that allow them to escape the police when crossing borders (Mogel and Bhagat 2007). Sometimes a tool in the service of a dominant power, sometimes a means of denouncing or circumventing it, the map can, in fact, be a fully political object (Stienne 2019).

1 Frontex is a European agency that is dedicated to managing operational cooperation at the external borders of the Member States of the European Union.

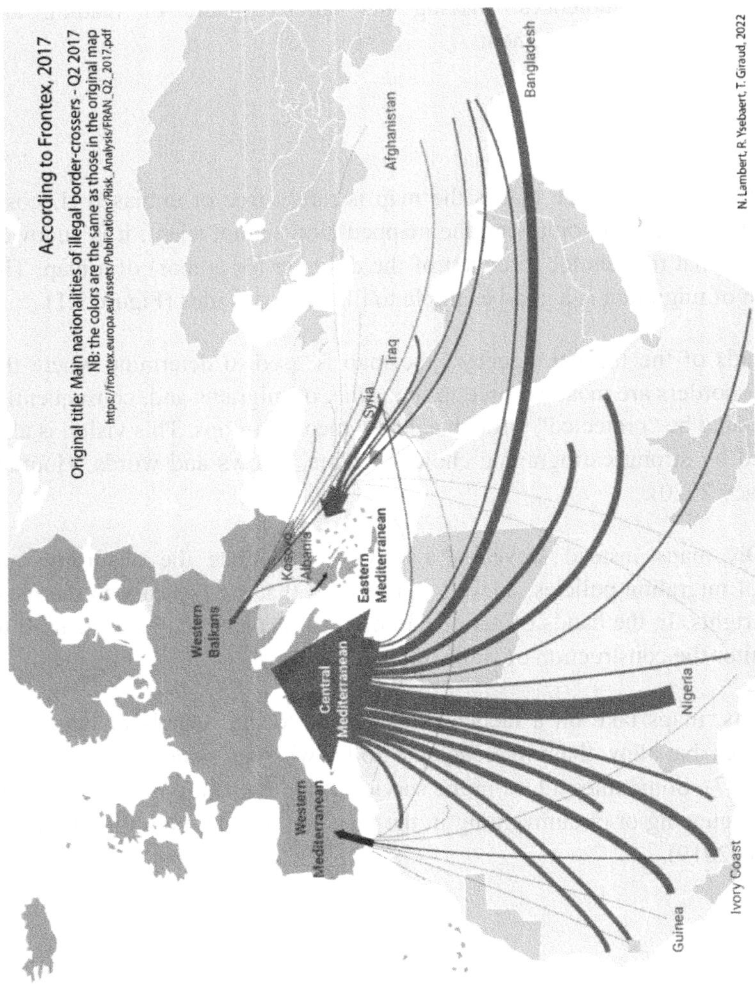

Figure 3.1. *The rhetoric of invasion according to Frontex. For a color version of this figure, see www.iste.co.uk/mericskay/cartographic.zip*

3.2.3. *Cartography: above all a matter of choice*

Cartography is a profoundly dialectical practice, in that it allows for demonstration, rebuttal, argument and reasoning. It is also both inherently scientific and objective, and subjective and biased. The subjectivity of maps, which geographers have been slow to assume as such, can be explained in several ways. First, no single map can claim to represent the full complexity of the real world. According to Goodman (1972), "there is no such thing as a completely adequate map, for inadequacy is intrinsic to cartography". It is a fact that clarity requires selection, simplification and generalization. You cannot put everything on a map. A good map will therefore be one that simplifies, summarizes, prioritizes, omits and orders, in short, one that achieves a schematization of reality that is essential for the emergence of keys to understanding, within the framework of a particular subject or objective. The map is used to "understand and search for what is going on behind simple appearance" (Pelletier 2013).

By reasoning through the absurd, many authors, from Carroll (1893) to Eco (1992) and Borges (1946), have digressed on what would be an unsimplified map (Palsky 1999), that is to say, a life-size map, which would represent everything in the world and would be superimposed on it at every point. Beyond the impossibility of such a production, it would be a very useless undertaking (see on this subject the tale *La Carte de l'Empire*, told by Chamussy (1982)). Simplification is an integral part of the cartographic process. Some authors even go so far as to say that maps lie, that they distort the truth to help the user see what the author wants him or her to see (Monmonier 1991). Ultimately, every map is the result of an intention, an intellectual vision, but above all, of choices that the cartographer must accept, endorse, discuss and explain.

3.2.4. *The cartographic language and its rhetoric*

In order to describe and propose representations of the world, cartographers have a varied graphic language that can be implemented with a palette of diversified tools, each with its advantages and disadvantages. Every map carries a message that is expressed through colors, lines, words, exaggerations and omissions (Lambert and Zanin 2020). There is a wide range of possibilities offered to cartographers: selection of data, choice of mesh and geographic space, statistical transformations, geometric transformations, graphic choices, colors, layout, projection, generalization, coverage, etc. At this point, it should be noted that although there is a strong link between the nature of the data used and the way it is represented, there is no mechanical relationship between the two. In other words, the same information can give rise to a multiplicity of representations (Zanin and Lambert 2012). The language of cartography is thus linked here to the art of narration of eloquence. There are facts and the way they are told. In cartography, there are data and the way in which they are mapped.

The fundamental concepts of rhetoric, which were defined by Greek philosophers more than 20 centuries ago, particularly by Aristotle, apply very well to contemporary cartography. They rest on three pillars: *logos*, *pathos* and *ethos* (Figure 3.2). The *logos* appeals to reason, to the intellect, to logical reasoning. It is the message itself, the content. As for the *pathos*, it refers to style, emotions and imagination. In cartography, for example, this will consist of choosing round and harmonious lines rather than sharp and abrupt angles, or arbitrating between favoring small signs in order to minimize the importance of a phenomenon or, on the contrary, magnifying them strongly, to the point of graphically saturating the image, to give it weight. Lastly, *ethos* refers to values and credibility. It determines the trust given to the producer of the map, the reputation of a researcher, the reputation of an institution, of a journal or the credibility of the data represented. The more *ethos* a map has, the more authority the arguments deployed on it have.

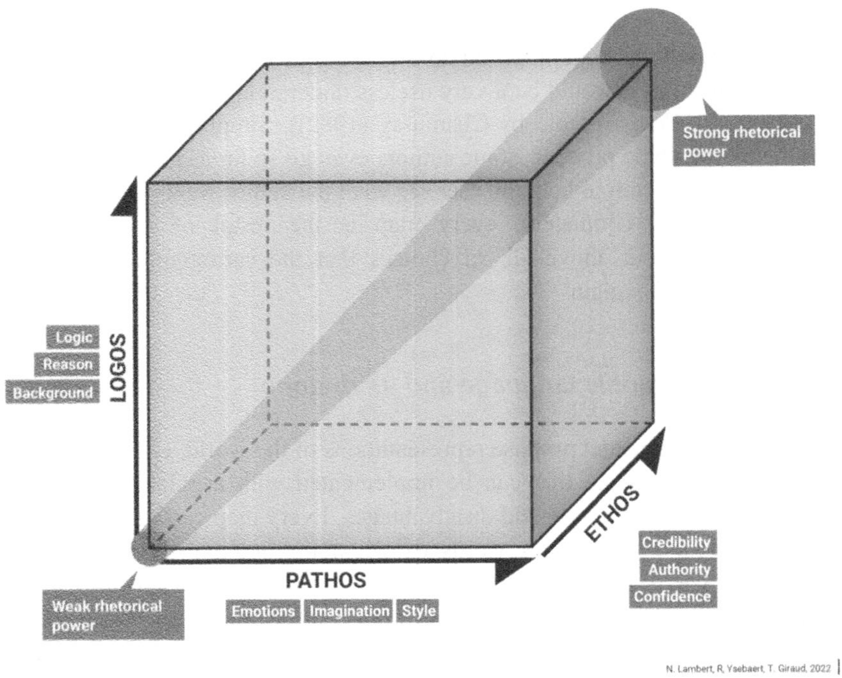

Figure 3.2. *The cube of cartographic rhetoric (Lambert and Ysebaert 2021). For a color version of this figure, see www.iste.co.uk/mericskay/cartographic.zip*

To understand the role of rhetoric in cartography, Bunge's work, and in particular his map of infants bitten by rats in Detroit's inner city in 1975, is a good example (Popelard and Vannier 2009). In constructing such a map (Figure 3.3), Bunge not only spatially represents a simple statistical data, he also offers a vision of it. Through a sharp and brutal detail (rat bites on infants), he reveals a social reality: squalor, precariousness, poverty, inequality, insecurity, etc.

However, this is a well-known rhetorical technique: using precise details and descriptions in a story or speech that are so vivid, that they force those that are listening to reconstruct the entire scene inwardly. Bunge distinguished between what he called skeletal maps, which represent statistical information, and life maps, which tell the story of people's lives, two complementary approaches according to him (Bunge et al. 1971).

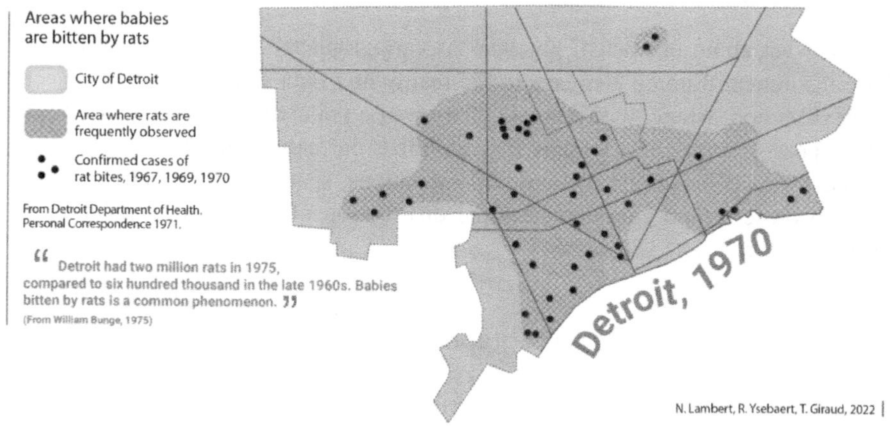

Figure 3.3. *Bunge's radical geography (from Lambert et al. 2021). For a color version of this figure, see www.iste.co.uk/mericskay/cartographic.zip*

In short, a map will be all the more effective if it is eloquent and capable of captivating a given audience (the construction of the map itself being thought out according to the audience for which it is intended) in order to transmit a maximum of information in a minimum of time. To do this, the art of cartography will consist of constructing a credible, coherent, speaking image, where all the elements of the map will be linked together according to a discursive logic that will speak to both reason and emotions (Migreurop and Clochard 2017).

3.2.5. *Traceable and reproducible transformations*

Insofar as maps "lie" (the first "lie" consists of passing from a sphere to a plane) and that they are the result of intentional communication processes, what about their scientific nature? It will always be difficult, if not impossible, to demonstrate the objective or neutral character of a map, which is why one must at least be able to guarantee its honesty. This requires, of course, a willingness to inform rather than to manipulate.

However, it also means considering cartographic creation as the end of a traceable documented transformation process (Tobler 1979). Making data accessible, communicating on processing, sharing codes and explaining the mapping process all contribute to making cartographic practices more transparent, as well as to guaranteeing a certain honesty from the cartographers and their productions.

But what about our actual practices as cartographers? How can we implement these theoretical framing elements to the reality of a data set? This is the subject of the next section, where, with a single data set of a population grid on a world scale, we propose a series of 13 maps that mobilize the cartographer's set of tools in various forms (filters, aggregations, discretizations, spatial transformations, etc.) in order to present (and communicate) the distribution of the world population in a varied manner.

3.3. Multi-representation cartography: variation on a theme

After a presentation of the data used and some methodological elements of framing, we explore and discuss here the discursive aspects related to the modes of cartographic representation.

3.3.1. *Presentation of the multi-representation approach*

Here, using examples, we propose demonstrating how a map is capable of saying much more than the simple raw data that was used to construct it. To do this, we take up Queneau's (1947) stylistic exercise, which has already been transposed to mapping the GDP of the world's countries (Lambert and Zanin 2016) and the people who died in migration in the Mediterranean (Giraud and Lambert 2019). The objective of such an approach is to propose a multiplicity of modes of cartographic representation built on identical data[2]. Like an author of literature, the cartographer must make choices. He must select information, sort it, prioritize it and select it in order to summarize it in an intelligible and meaningful way, while simultaneously representing it in a clear and efficient manner.

3.3.1.1. *Technical implementation*

As part of this approach, we have chosen to use the programming language and free software R (R Core Team 2020) to produce all the maps presented in this chapter. This technical choice is justified by the quality of the language's spatial ecosystem (Bivand 2020), particularly in terms of scalability and functionality. A large number of extensions (packages) allow for both the implementation of a wide range of spatial analyses and the production of a large variety of cartographic representations. Moreover, the possibility of executing almost all processing chains (acquisition, formatting, processing and representation of data) within the same environment avoids software disruption and ultimately facilitates the reproducibility of the approach and results (Giraud and Lambert 2017).

NOTE.– All the maps presented in this chapter are available online in high definition, along with the source code that allowed them to be generated. With a view to transparency, the availability of the processing chains (in the form of scripts) makes it possible to increase trust in our productions by opening the cartographic production's "black box" (Nüst and Pebesma 2021).

Figure 3.4 describes where our approach falls on a spectrum of cartographic reproducibility (Peng 2011). In this approach, data formatting, processing and cartographic representation were carried out because of programs in the R language. The latter use the reference R libraries to format and manipulate spatial data (*raster*, *sf*), then to map them (*cartography*). Some more specific libraries are also used for more advanced processes (Stewart's potential calculation with *SpatialPosition*) or specific modeling (*tanaka*, *linemap*). This part of the work is therefore perfectly traceable and reproducible, as shown on the website linked to this chapter[3], which provides the full code used to generate the cartographic representations presented here.

On the other hand, it is often necessary to use vector graphics software to complete their layout: the final presentation of these maps (hierarchy of texts,

2 For more information on this matter, see Chapter 1, which discusses this type of approach in detail, and Chapter 4, which presents an example of multi-representation mapping (of flows) around the theme of global container trade.

3 www.rcarto.gitpages.huma-num.fr/multirepresentation/.

positioning of labels, design of legends, etc.), which is more a matter of graphic design, was carried out with CAD software[4], where techniques can certainly be documented, but are not reproducible.

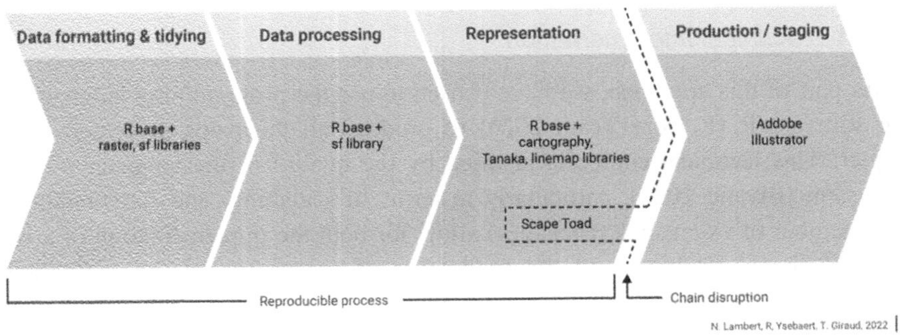

Figure 3.4. *Spectrum of reproducibility in cartography. For a color version of this figure, see www.iste.co.uk/mericskay/cartographic.zip*

3.3.1.2. *Input data*

All of the following cartographic representations originate from a single data set, The Gridded Population of the World (GPW)[5]. Among the 10 or so indicators made available by the CIESIN (Center for International Earth Science Information Network), we will choose the simplest and most up-to-date: the total population in 2020.

The demographic indicators proposed by the CIESIN are the result of a very large data collection carried out at the most detailed administrative level for all countries in the world based on national censuses between 2005 and 2014. These statistical data are then matched to the reference geometries of each country, and an estimate is made for the year 2020 by following the trends observed in the last two censuses, along with the United Nations estimates, before being broken down into a raster model (GeoTIFF) at different levels of granularity. The finest resolution available corresponds to 30 s (about 1 km at the equator) and the largest to 60 min (about 110 km at the equator). For this exercise, the 30-min resolution was chosen[6].

4 Computer-aided design.

5 GPW is a raster grid produced by CIESIN (Center for International Earth Science Information Network, Columbia University) that models human population distribution (counts and densities).

6 As a result of the distortions introduced by the projections, the resolution of the mesh is equivalent to 55 km at the equator, 35 km at intermediate latitudes (Paris), and less than 10 km north of Greenland.

At this level of spatial resolution, there are 70,123 cells covering land areas worldwide (Table 3.1).

2020 cell population	Occurrences	Frequency (%)
Less than 1 inhabitant	11,075	15.8
1–10 inhabitants	2,184	3.1
10–100	7,247	10.4
100–1,000	8,973	12.8
1,000–10,000	13,550	19.3
10,000–100,000	15,361	21.9
100,000–1 000,000	10,028	14.3
1,000,000–23,065,668	1,705	2.4
Total	70,123	100

Table 3.1. *Population interval counts (logarithmic scale) of CIESIN grid cells (resolution = 30 min)*

This great disparity in the distribution of populations within the cells is also observed in terms of cumulative frequency (Figure 3.5). We can see that 50% of the world's population (more than 4 billion people) is only concentrated in 2.4% of the 70,123 cells in the data set. Conversely, 50% of the cells in the data set have less than 14 million inhabitants (0.2% of the world's population).

The distribution of this statistical variable, which is very asymmetrical on the left (the median is equivalent to a cell with 2,708 inhabitants, compared to 110,600 for the mean), reveals useful possibilities for the cartographer who wishes to graphically transcribe the specificities of this data set. How can the maximum population concentrations contained in certain cells, or certain portions, of the world space be rendered? How are "voids", which occupy most of the planet, described?

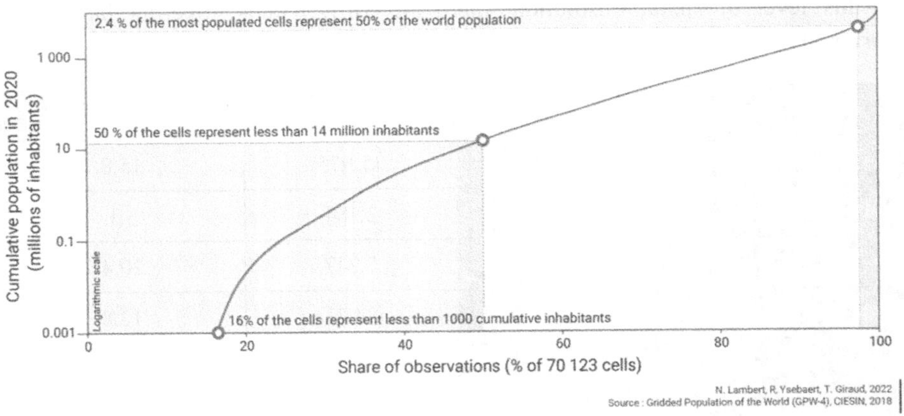

Figure 3.5. *Cumulative frequency of populations within the grid*

This short description of the data set used also reminds us of some important things to keep in mind before embarking on any cartographic production. Like maps, data are sociotechnical constructs (Chrisman 2005). Their collection, processing, updating, and distribution have costs, as well as political, social, or economic objectives, and there is always a bias toward the fact that they reflect an objective reality. In this sense, one of the most striking case studies of population data is provided by the example of the Nigerian census where, after several censuses carried out in chaotic conditions, this country, which was the most populated in Africa, lost 30 million inhabitants overnight in 1991 (Lévy and Omoluabi 1992).

The cartographer, being a major user and "manipulator" of heterogeneous data, must therefore be particularly vigilant with regard to the sources used, and measure the consequences that the latter can induce and involve in cartographic communication. Indeed, data used with poorly controlled or poorly informed sources can undoubtedly introduce cartographic lies and undermine the credibility of the message.

3.3.2. *Cartographic constants*

As previously mentioned, all maps lie, and some lies are constants in any cartographic construction.

3.3.2.1. *Cartographic projection(s)*

The first lie is representing a spherical world in three dimensions on a flat surface in two dimensions. This criticism is not new. In the 19th century, Reclus already considered that teaching geography using false images was total nonsense (Reclus 1903). However, because of their capacity to distort, cartographic projections are, in their diversity, precious weapons in the hands of cartographers, as distorting the world conveys in itself a vision of the world. Beyond the matter of geodesic accuracy, many map projections are, in practice, designed for the message they convey (Figure 3.6). The message of each map is always the result of a complex composition associating content (the statistical data represented) and a container, the "setting" on which it is based (Rekacewicz 2014). Each map is therefore a production that is reminiscent of the world of theater.

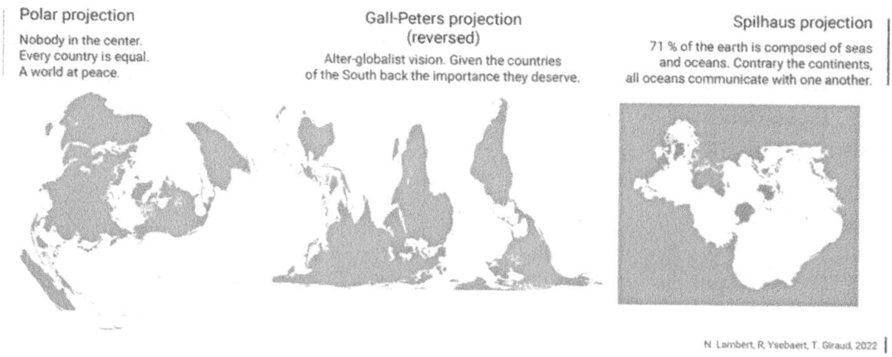

Figure 3.6. *Each projection has its own message (Lambert et al. 2021)*

The Gall-Peters projection, for example, which shows the real surface area of the continents to the detriment of their shape, aims to rehabilitate the countries of the South, which are crushed on traditional cartographic representations by a dominant North. The polar projection, used for the logo of the United Nations, which does not put any country at the center, came about after the Second World War in order to establish a desire for peace and equality. As for the cartographic projection of the oceanographer Athelstan Spilhauss, which is very original, it only aims to draw attention to the importance of the seas and oceans on the surface of the Earth. Much more than simple media for geographic information, the various cartographic projections are vectors of discursive elements, which carry a narrative, and are an integral part of the cartographic message.

Although we are aware of the extent of the rhetorical possibilities of cartographic projections, we have considered that they are outside the scope of our study. In the demonstration to follow, we will reason with a constant projection and therefore a constant "setting", that is, a world distorted by an equidistant cylindrical projection that is classically centered on Europe. This choice of a regular representation of the world, one that is as neutral as possible, aims at emphasizing the cartographic representations of statistical data.

3.3.2.2. *Cartographic scale(s)*

For the same reasons, we will work on a constant scale, that is, the world. The geographical breakdown and the scale of analysis have a significant influence on the modeled results. Each scale of analysis or observation brings its own keys for reading (Figure 3.7). "Sometimes, we have to look at the earth through a microscope and sometimes from the top of a satellite [...] Reality appears to be different according to the scale of the maps, according to the levels of analysis" (Lacoste 1976). In the cartographic style exercise presented here, we will therefore only work on a global scale, with a constant cartographic projection and level of geographical regularity, using data made available by CIESIN at a resolution of 30 min. In some cases, these data can be aggregated into grids of different sizes or administrative meshes.

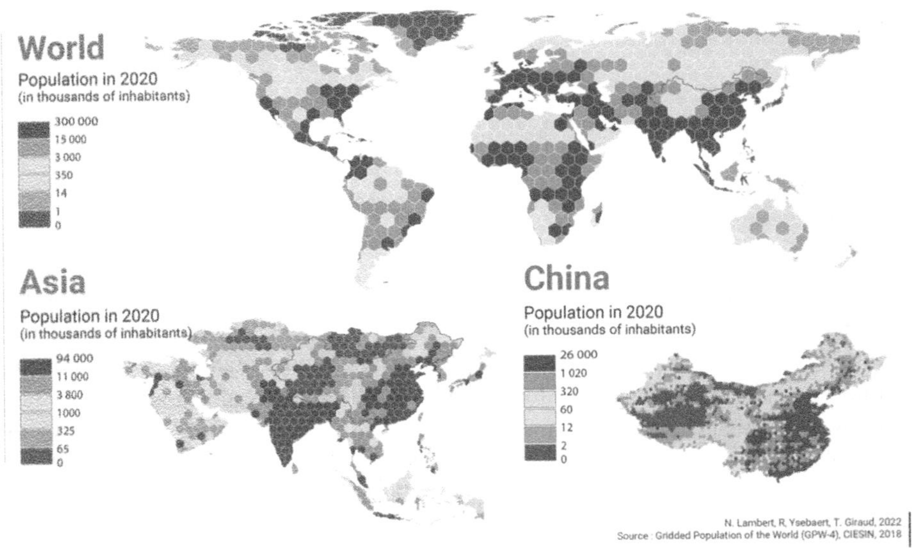

Figure 3.7. *"Reality" appears differently depending on the scale (Lambert et al. 2021). For a color version of this figure, see www.iste.co.uk/mericskay/cartographic.zip*

3.3.3. *Discourse and mode of cartographic representation*

The choice to represent relative or absolute data is a choice that is fraught with meaning, and is assessed in relation to the theme under consideration. Be it the standard of living or economic power, it all depends on what you want to show. Mapping absolute or relative data means referring to different realities and, above all, communicating distinct, divergent and even contradictory messages. Moving from the absolute to the relative can indeed be a way of minimizing and relativizing a phenomenon, or, on the contrary, giving it a more social aspect. To go from the relative to the absolute is to remember that, in power relationships, it is always the number that counts.

When applied to the world population, the cartographic representation of absolute data by proportional signs (Figure 3.8) refers to the demographic weight of states, while the representation of population densities by solid colors (Figure 3.9) focuses the cartographic message on the unequal distribution of humans on Earth and thus highlights the small, densely populated countries (Monaco, Malta and Bahrain).

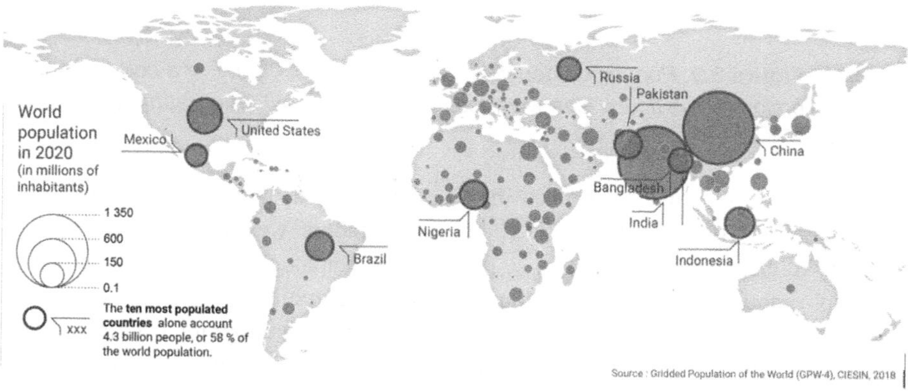

Figure 3.8. *The unequal demographic power of nation-states (Lambert et al. 2021). For a color version of this figure, see www.iste.co.uk/mericskay/cartographic.zip*

On proportional maps, graphic semiology allows one to "free oneself from the framework" imposed by the base map, giving all its place to the signs. By moving from a zonal implementation (countries) to a punctual figuration (signs), the cartographer "plays tricks" by imposing figures on the map that more or less overflow from the strict place they designate, or on the contrary, only partially cover it (Brunet 1987), thus making it possible to best express the represented proportions. However, choropleth maps, constructed on discretized relative data, only allow the expression of a hierarchical relationship between places, and the choice of class limits (discretization) can strongly influence the message. Although these two methods are useful and complementary for analyzing and representing a geographic space, they deliver, by construction, messages of a radically different nature.

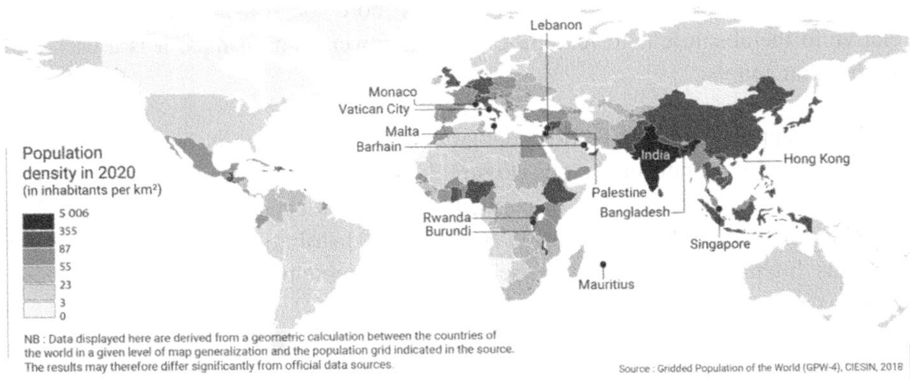

Figure 3.9. *Population densities (Lambert et al. 2021). For a color version of this figure, see www.iste.co.uk/mericskay/cartographic.zip*

This idea of power can also be conveyed in a number of cartographic transformations (Cauvin 1998), including cartographic weight transformations (a category of anamorphosis), that go so far as to emancipate themselves from physical geography and transform the contours of the geographic objects that serve as a medium for the cartographic representation. Here, the cartographic deformations performed according to the method developed by Gastner and Newman in 2004 (Gastner and Newman 2004), and implemented in the ScapeToad software (Andrieu et al. 2008), modify the size of each pixel according to the population value associated with it, before grouping them by country, in order to reform an alternative territorial mesh (Figure 3.10).

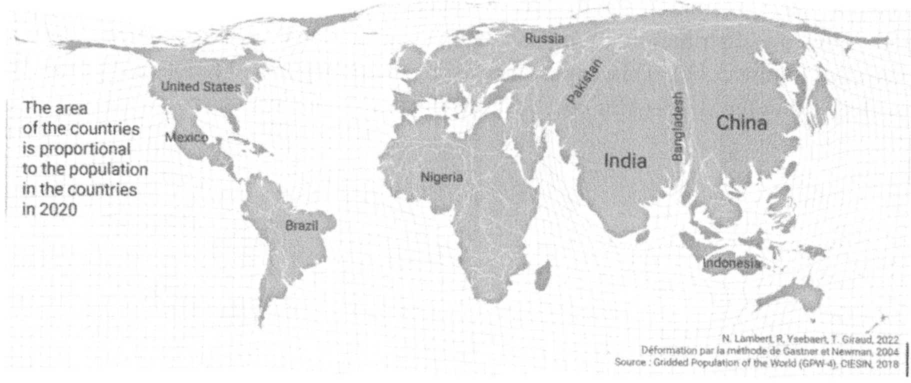

Figure 3.10. *Demographic distortion (Lambert et al. 2021)*

On this type of map, the great demographic powers take their place beyond their own "borders", while the least populated countries shrink until they sometimes disappear from the surface of the map. The use of this new base map stripped of its geographical "reality", thus produces a true rhetoric of power relations with winners and losers.

3.3.4. *Breaking out of the geometric constraint*

Far from these images that exacerbate the grandeur and omnipresence of nation-states, and even though statistical data are often provided by these states themselves and one of the historical correlates of nation-states is statistics (Grataloup 2009), other cartographic representations aim to free themselves from territorial divisions. Some transformation methods allow for the disaggregation or interpolation of data in order to represent them in a way that is different from the collection grids and administrative divisions. While this has major methodological value in overcoming the statistical and cognitive difficulties associated with mesh heterogeneity (Openshow 1979; Grasland et al. 2006), it also allows for the simplification of spatial structures by showing geographic space in "natural" or continuous logics (Commenges et al. 2015). In other words, it allows us to propose worldviews that are no longer international but global (Grataloup 2018; Capdepuy 2020).

The oldest method of disaggregating data is probably the dot map, the first known realization being that of Armand Joseph Frère de Montizon in 1830, on which one dot represented 10,000 people (Palsky 1984). This type of map, which is very educational, is widely used in school textbooks to describe settlement areas.

The dot maps give the image of a representation as close to the ground as possible, and show the distribution of human beings on a global scale as objectively as possible (Figure 3.11). This appearance is all the stronger if the dots are precisely located within the states (near coasts, large cities, communication routes, etc.).

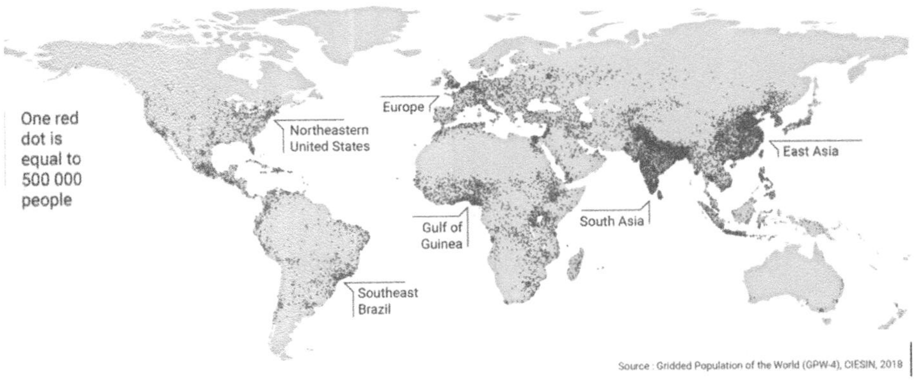

Figure 3.11. *World population in 2020 (Lambert et al. 2021). For a color version of this figure, see www.iste.co.uk/mericskay/cartographic.zip*

However, we must not lose sight of the fact that we are dealing with an image constructed from scratch, whose underlying technical processes are complex. Here, for example, the dots are distributed within a grid using a random method. The finer the grid, the more the visual rendering gives the impression of an image that is close to reality, hence the interest in documenting the methods to make the construction of this type of map transparent and understandable.

A variant of this dot map representation of the world population consists of spreading proportional symbols (often circles) evenly over the surface of the map and modifying the surface according to the underlying data. For greater legibility, these signs can be moved away from each other (Dorling 1996) until they fall outside the initial grid. This is an interesting technique to avoid the superimposition of cartographic figures on each other (classical proportional circle maps), as well as to show the whole of the masses present on a given geographical space, in a synthetic way. Symbolically, like cartograms, this representation can evoke the idea of overflow (Figure 3.12). In our example, regional powers (and no longer state powers) emerge (South Asia, East Asia, the Gulf of Guinea), while empty spaces appear in the background.

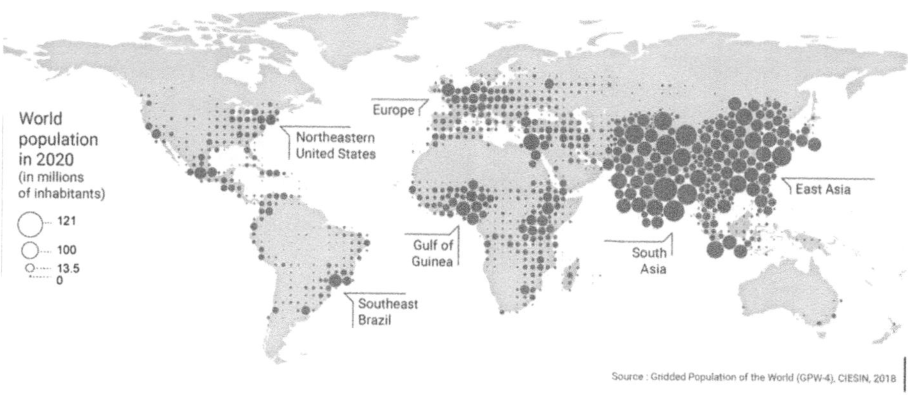

Figure 3.12. *Demographic overflow (Lambert et al. 2021). For a color version of this figure, see www.iste.co.uk/mericskay/cartographic.zip*

To get rid of the boundaries, the methods of smoothing and spatial interpolation offer other perspectives (Stewart 1942). By calculating the masses in a given geographic neighborhood, whose size decreases with distance (calculation of potentials[7]), and then extracting isolines and isosurfaces of comparable values, these methods make it possible to draw a world of continuums that are independent of borders, far from the discontinuous logic of maps of nation-states (Figure 3.13).

To produce such a map is to propose a world of seamless gradients, where the data are established from one end of the globe to the other. However, others may also see it (ESPON 2012) as an indirect way of discussing “demographic reliefs”, where the demographic chain of the Punjab rubs shoulders with the coastal peaks of the China Sea, where the Maghrebian cuesta front plunges into the vast Saharan depression, or where the immense Latin American atoll shows a few demographic resurgences in the center of the Amazonian forest.

7 This calculation is based on a mathematical function that decreases with distance, used to estimate the population mass in a geographic neighborhood (200 km here). The basic parameters used in this calculation (function, impedance of the function and distance) are captioned in the map in Figure 3.13. Their correct use, based on assumptions of interaction between places, is essential for the controlled construction of this type of cartographic representation.

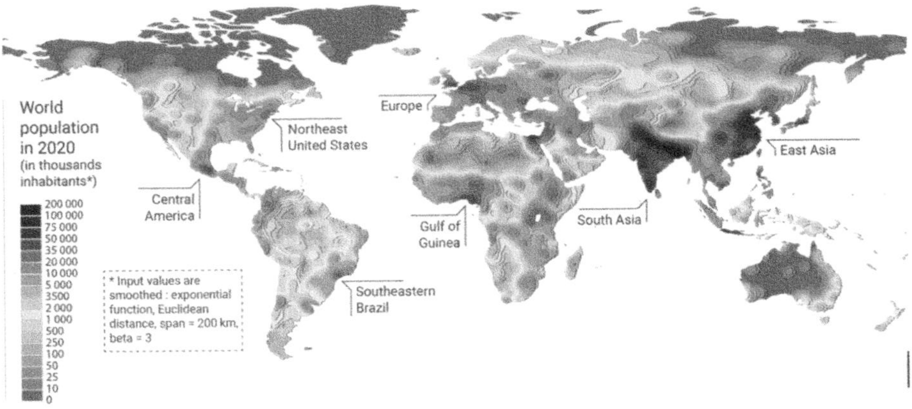

Figure 3.13. *A world without borders (Lambert et al. 2021). For a color version of this figure, see www.iste.co.uk/mericskay/cartographic.zip*

3.3.5. *The public challenge*

Whatever the method used, no map can be constructed without considering the audience for which it is intended. Children, adults, specialists, decision-makers and scientists are not used to the same cartographic images or to the same complexity of the underlying discourse.

Thus, in a willingness to raise awareness and even convince as many people as possible, a certain number of extremely simplified maps are appearing on the Internet and social networks, with the aim of being understood in an extraordinarily short time (to the detriment of the real complexity of the phenomena represented)[8]. This new cartographic genre (Schweikart and Domnick 2013) is closer to infographics, with key figures highlighted, than to the traditional constructions of thematic cartography where spatial structures prevail. On these representations, the message must be understood at a glance. No time for thinking, what counts is the number of views, the sharing, the audience, the reactions, and especially the speed of diffusion of the map on the web (Robinson 2018).

Summarizing the complexity of a phenomenon with a number or a striking image is not an easy task, but is particularly effective when the approach is well done (the progress of urbanization in soccer fields, the ecological footprint expressed in number of planets, etc.). Summarizing an inextricable and complex

8 For more information on the challenges of cartographic communication on the web and social networks, see Chapter 2 and Chapter 6.

phenomenon into a punchline makes it possible to reach the largest number of people. This logic also applies to cartography: many maps circulate, and are widely relayed and commented on through the web and social networks, because their message is unusual and/or understood in an extremely short period of time. The two maps below (Figures 3.14 and 3.15) are examples of this. They are not of major geographical interest, but this type of representation responds to an ever-increasing demand from the general public for factual elements to understand the world, and have it represented in a simple and effective way. However, it is important that oversimplification does not introduce erroneous information.

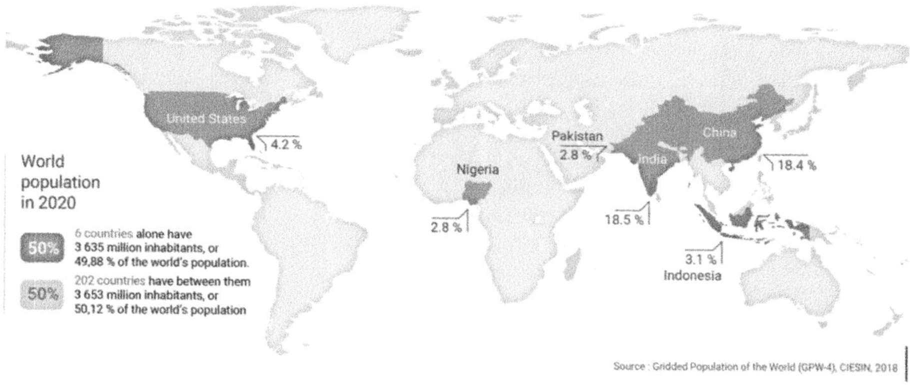

Figure 3.14. *Half of the population lives in just six countries (Lambert et al. 2021). For a color version of this figure, see www.iste.co.uk/mericskay/cartographic.zip*

Figure 3.15. *Half of the world's population lives on 3% of the earth's surface (Lambert et al. 2021)*

3.3.6. *Cartographic representations for specific themes*

As one explores the diversity of maps made around the same data set, it becomes clear that the cartographic possibilities for expressing and communicating ideas, and even intentions, are almost infinite. Sometimes, in order to make the cartographic narrative coherent, one will even seek to use a technical process that echoes the theme being addressed. The map below (Figure 3.16) takes up the principle of the first law of geography, put forward by Tobler (1970), for whom "everything interacts with everything, but two objects close to each other are more likely to do so than two objects far away".

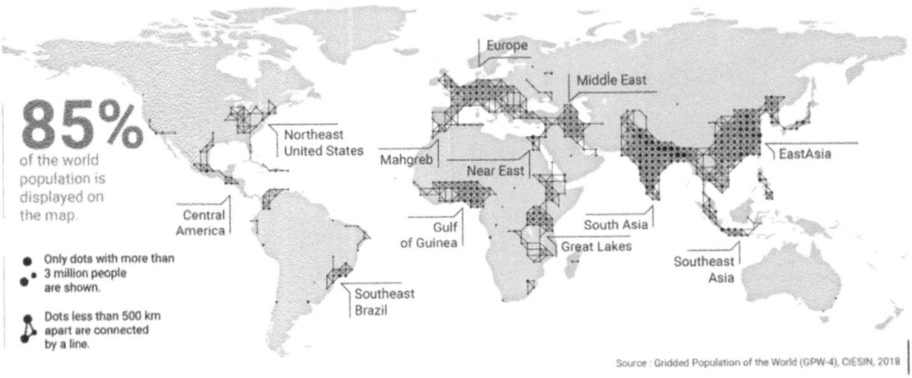

Figure 3.16. *Everything interacts with everything (Lambert et al. 2021)*

This idea has been taken up in Figure 3.16, showing a system of lines that connects places with more than 3 million inhabitants that are within 500 km of each other (according to a Euclidean distance). This threshold, although arbitrary, makes it possible to account for an interconnected world where two nearby places are likely to interact more than two distant places. Furthermore, the lines are a clever way to highlight and visualize regions that are fairly dense and close to each other. Note that this map could have been produced with a different distance threshold and a different, more or less fine spatial aggregation.

As we can see, linking the method of representation with the message can be very effective. So, how do you express the fact that more than half of the world's population now lives in cities? The solution that was proposed here was to symbolize the data in three dimensions and to work on height (Franklin and Lewis 1978). The urban population centers are thus represented by extruded hexagonal meshes symbolizing buildings, whose height varies according to the number of inhabitants (Figure 3.17). The heavy urbanization of certain areas of the world is thus materialized on the map by a series of urban columns (skyline), where both the height and density of the buildings are mixed.

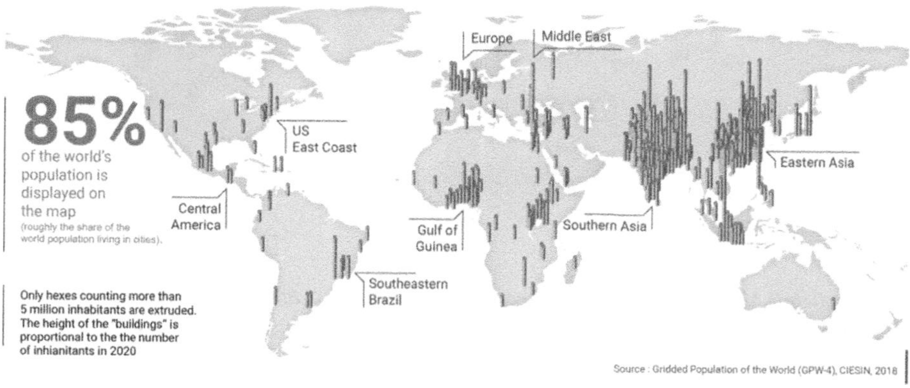

Figure 3.17. *Homo urbanus (Lambert et al. 2021). For a color version of this figure, see www.iste.co.uk/mericskay/cartographic.zip*

Through the use of graphic semiology, statistics or geoprocessing, there are a thousand and one ways to make geographic data speak and transform into graphic discourse. Beyond the aspects that are strictly related to graphic semiology as a whole, only selecting data near seas and oceans, for example, allows us to reveal that almost half of the world's population lives near a coast (Figure 3.18). By focusing on this aspect, the map takes on a political aspect, since according to the IPCC, water levels could rise by more than 1 m by the end of the century.

In the same way, focusing on places that are relatively populated in terms of their neighborhood (the resolution and definition of the neighborhood naturally being configurable) instead of on places that are populated in absolute terms, makes it possible to show a map in which the highlighted areas are not necessarily those where the inhabitants are the most numerous, or even where the population density is the highest (Grasland 1999; Dubois et al. 2007; de Ruffray et al. 2011). Such a map (Figure 3.19), which could be characterized as "local" map, gives a completely different image of the world, a "glocal"[9] image that highlights local potential.

9 The term "glocal" is a catch-all word that means both local and global. It was popularized in the English-speaking world in the 1990s by sociologist Roland Robertson.

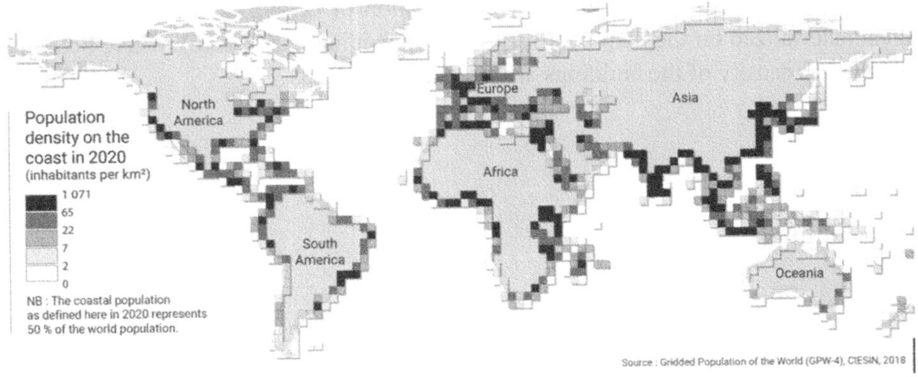

Figure 3.18. *Humanity in the face of rising oceans (Lambert et al. 2021). For a color version of this figure, see www.iste.co.uk/mericskay/cartographic.zip*

This representation presents a vision where isolated poles emerge and coalesce with the major globalized metropolitan backbones. It also highlights an unusual network of agglomerations, specifically linked to pioneer fronts and historical routes, whether in Russia (along the Trans-Siberian railway), Amazonia (along the Amazon passing via Manaus) or western China (Silk Road).

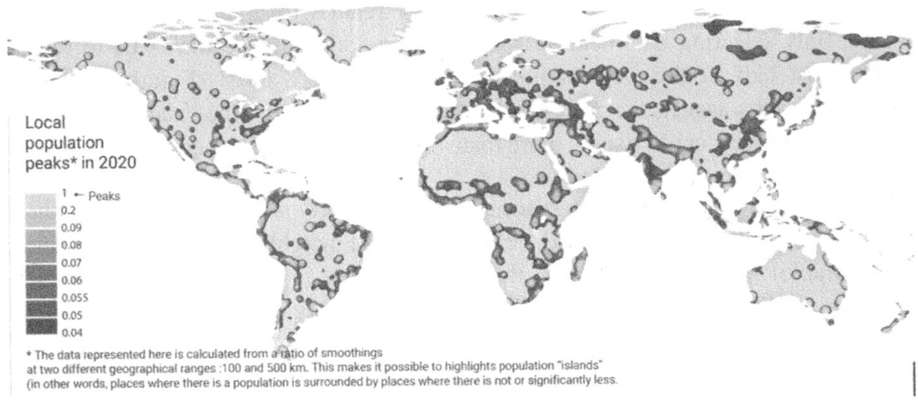

Figure 3.19. *Glocalization (Lambert et al. 2021). For a color version of this figure, see www.iste.co.uk/mericskay/cartographic.zip*

Lastly, maps can also be invitations to dreams and adventure. Blank spaces on maps have always been a dream. Unknown lands open up prospects for adventure, which many geographers are fond of. This map of demographic voids (Figure 3.20) reflects this idea. By focusing representation on distant and unknown lands (Tiberghien 2007), it invites adventure. Obviously, other representations could have been used to visualize this theme such as anti-cartograms, which consist of inverting the representation of solids and voids (Poncet 2011).

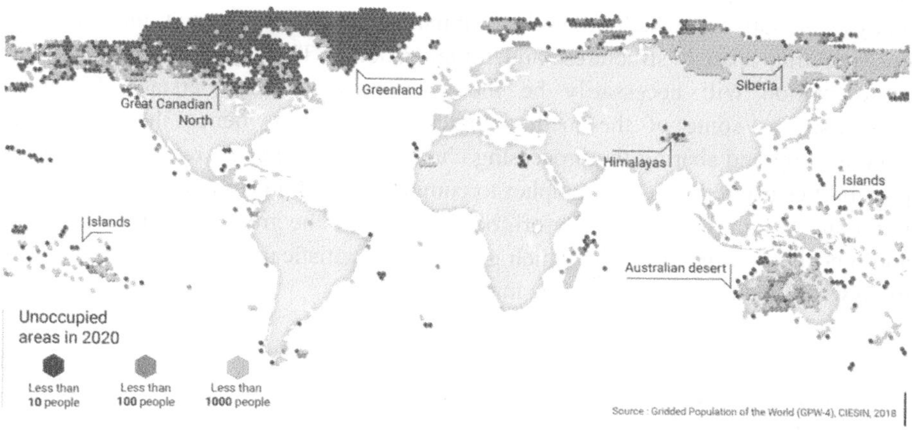

Figure 3.20. *Terrae incognitae (Lambert et al. 2021). For a color version of this figure, see www.iste.co.uk/mericskay/cartographic.zip*

3.4. Conclusion

This exercise of multi-representations, which has been materialized by 13 diversified maps, illustrates that from the same information, with a constant projection, scale and coverage, it is possible to build a vast range of cartographic proposals, each one being built according to its own method and conveying its own message. From a data set of world populations, we have thus highlighted some masses and hierarchies, and some spatial concentrations (peaks, exceptional values), as well as continuums and disruptions in space.

A possible summary of this palette of spatial representations consists of cross-referencing the modes of construction and the objectives of communication to reveal cartographic intentions (Figure 3.21). We understand the "cartographer's intention" here as the methodological and graphic choices that is inherent in cartographic construction, with the aim of communicating a message to a given audience. Depending on the mode of representation, the reader will not acquire the same knowledge about the distribution of the world's population.

In their diversity, these maps can thus be educational and accessible: the geographic information has been filtered, purified and contextualized to focus on certain aspects of the data set. They are able to be "conventional" and to mobilize characteristic visual variables of graphic semiology. These maps are also likely to be based on geographical distortions that allow the cartographer's proposed approach to be supported graphically. Lastly, they can be based on methods derived from spatial analysis or statistics, which allow the original data to be transformed, thus bringing out specific spatial structures and organizations. This type of cartographic representation will necessarily be more complicated to understand for non-specialists, as some of the methodologies are complex, hence the interest in documenting and sharing the processing chains. In any case, the range of graphical possibilities offered by a cartographer to communicate to a given public what seems important to him or her will, a priori, be all the wider the greater his or her mastery and hindsight on the methods (which graphical or statistical process should be used to do what) and tools used (how the map is made).

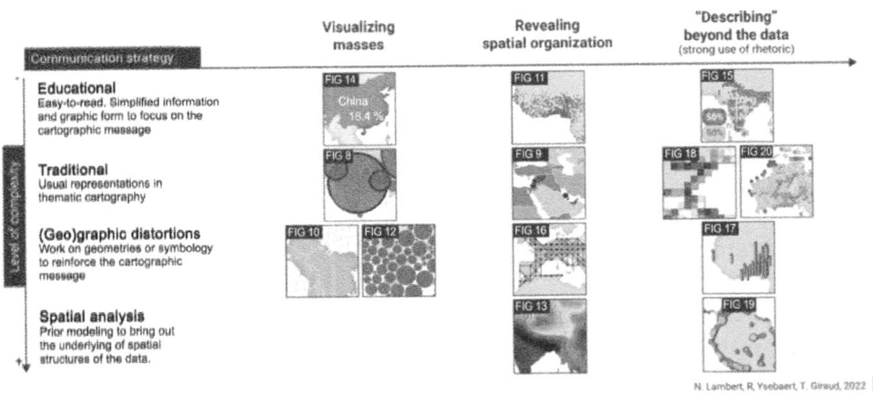

Figure 3.21. *Summary of the multi-representation exercise. For a color version of this figure, see www.iste.co.uk/mericskay/cartographic.zip*

Beyond this inevitably improvable summary proposal, we wanted to first of all demonstrate that there is no mechanical and automatic relationship between geographic data and their cartographic translation, and beyond the technical choices, any map essentially results from an intention from a communication strategy. We therefore do not claim, at the end of this stylistic exercise, to have exhausted the subject; it is quite the contrary. In fact, statistics and spatial analysis allow for a whole range of representations. By taking distance, surfaces, topology and underlying meshes into account to transform data, the map can take on extremely diverse forms (continuums, disruptions, multiscalar representations, etc.) to convey various messages.

One of the keys to the originality and effectiveness of maps is therefore not only in the semiotics of geographic data, but also in the preparation and analysis of these data, which implies cartographic intention. Making a communication map is not only limited to the "simple" formatting of spatial data, but today refers to broad skills in terms of data processing. This highlights the strong methodological aspect that is intimately associated with thematic cartography. This is especially true in the face of contemporary instrumentation, such as the R language and its multitude of packages, which offers numerous ever-changing possibilities, both for cartographic representation and spatial analysis.

Figure 3.22. *There is art in cARTography (Lambert et al. 2021)*

Other cartographic facets could have been explored and constructed in this sense. Here, we are thinking of the example of the exploitation of the third dimension (Grasland and Madelin 2001) or the possibilities of interactivity and animation made available today by online cartography[10]. We thus support the idea that the choices of representation go far beyond what has been shown here. No single map will summarize and exhaust a theme or even data. On the contrary, it is through a multiplicity of spatial representations to be compared, discussed and criticized that it

10 For more on the questions guiding this research, see Chapter 6 and Chapter 7.

is possible to grasp the complexity of a geographical space, as well as its contradictions.

Ultimately, we must not lose sight of the fact that maps have a great capacity to appeal to us and that their aesthetics sometimes border on the world of art, as this last map (Figure 3.22), largely inspired by the one imagined by Tobler in 1970, reminds us. The appeal of maps, a fully-fledged weapon of war in terms of communication, even manipulation, implies that cartographers must always reflect, experiment, rethink and question their productions.

3.5. References

Andrieu, D., Kaiser, C., Ourednik, A. (2008). ScapeToad [Online]. Available at: http://scapetoad.choros.place/ [Accessed 13 April 2021].

Bahoken, F. (2020). La carte thématique, objet scientifique source de manipulations. *Cause commune, production de territoires*, 14/15 [Online]. Available at: https://www.causecommune-larevue.fr/la_carte_thematique_objet_scientifique_source_de_manipulations

Bivand, R.S. (2020). Progress in the R ecosystem for representing and handling spatial data. *Journal of Geographical Systems* [Online]. Available at: https://doi.org/10.1007/s10109-020-00336-0.

Brunet, R. (1987). *La carte, mode d'emploi*. Fayard-Reclus, Montpellier.

Bunge, W., Heynen, N., Barnes, T. (1971). *Fitzgerald: Geography of a Revolution*. University of Georgia Press, Athens.

Cameron, D. and Jones, I.G. (1983). John Snow, the broad street pump and modern epidemiology. *International Journal of Epidemiology*, 12(4), 393–396.

Capdepuy, V. (2020). *50 Histoires de Mondialisations*. Alma Éditeur, Paris.

Cauvin, C. (1998). Des transformations cartographiques. *Mappemonde*, 49, 12–15.

Chamussy, H. (1982). La carte de l'Empire. *Brouillons Dupont*, 9, 51–59.

Chrisman, N. (2005). Full circle: More than just social implications of GIS. *Cartographica: The International Journal for Geographic Information and Geovisualization*, 40(4), 23–35.

Commenges, H., Giraud, T., Lambert, N. (2015). Enhancing spatial patterns representations by functional semantic simplification: An interactive tool for European stakeholders. In *27th International Cartographic Conference*. Rio de Janeiro.

De Ruffray, S. and Hamez, G. (eds) (2011). Enjeux des territoires transfrontaliers à l'échelle nationale. Vers la détermination et la délimitation de pôles transfrontaliers. Report, DATAR, RIATE.

Dorling, D. (1996). Area cartograms: Their use and creation. *Concepts and Techniques, Modern Geography (CATMOG)*, 59, 1–6.

Dubois, A. and Gloersen, E. (2007). Les disparités régionales et la cohésion : quelles stratégies pour l'avenir. Report, Parlement Européen, Brussels.

ESPON (2012). Europe's neighbourhood from a territorial perspective. Report, ESPON Internal Seminar, Paphos.

Franklin, R. and Lewis, H. (1978). 3-D graphic display of discrete spatial data by prism maps. *ACM SIGGRAPH Computer Graphics*, 12(3), 70–75.

Gastner, M.T. and Newman, M. (2004). Diffusion-based method for producing density-equalizing maps. *Proceedings of the National Academy of Sciences of the United States of America*, 101(20), 7499–7504.

Giraud, T. and Lambert, N. (2017). Reproducible cartography. In *Cartography and GIScience*, Peterson, M. (ed.). Springer, Berlin.

Giraud, T. and Lambert, N. (2019). Reproducible workflow for cartography – Migrants deaths in the Mediterranean. *Proceedings of the ICA*, 2, 38–44.

Goodman, N. (1972). *Problems and Projects*. Hackett Publishing Company, Indianapolis.

Grasland, C. (1999). Lissage cartographique et animation spatio-temporelle : quelques réflexions théoriques et méthodologiques. *Travaux de l'Institut de Géographie de Reims*, 101(104), 83–104.

Grasland, C. and Madelin, M. (2001). The unequal distribution of population and wealth in the world. *Population et Sociétés*, 368, 14.

Grasland, C. and Madelin, M. (eds) (2006). The modifiable area unit problem. Report, ESPON 3.4.3, Paris.

Grataloup, C. (2009). *L'invention des continents : comment l'Europe a découpé le monde*. Larousse, Paris.

Grataloup, C. (2018). *Vision(s) du Monde : Histoire critique des représentations de L'Humanité*. Armand Colin, Paris.

Ferland, Y. (2000). Les défis théoriques posés à la cartographie mènent à la cognition. *Cybergeo: European Journal of Geography* [Online]. Available at: http://journals.openedition.org/cybergeo/499

Houtum, H.V. and Bueno Lacy, R. (2020). The migration map trap. On the invasion arrows in the cartography of migration. *Mobilities*, 15(2), 196–219.

Jacob, C. (1992). *L'empire des cartes. Approche théorique de la cartographie à travers l'histoire*. Albin Michel, Paris.

Koch, T. and Denike, K. (2009). Crediting his critics' concerns: Remaking John Snow's map of Broad Street cholera, 1854. *Social Science & Medicine*, 69(8), 1246–1251.

Krakauer, J. (1996). *Into the Wild*. Presses de la Cité, Paris.

Lacoste, Y. (1976). *La géographie, ça sert, d'abord, à faire la guerre*. Petite Collection Maspero, Paris.

Lambert, N. and Zanin, C. (2016). *Manuel de cartographie : principes, méthodes, applications*. Armand Colin, Paris.

Lambert, N. and Zanin, C. (2020). *Practical Handbook of Thematic Cartography: Principles, Methods, and Applications*. CRC Press, Boca Raton.

Lévy, J.L. and Omoluabi, E. (1992). Le recensement du Nigeria. *Populations et Sociétés*, 272, 426–442.

Migreurop, and Clochard, O. (2017). *Atlas des migrants en Europe : approches critiques des politiques migratoires*. Armand Colin, Paris.

Miller, H.J. (2004). Tobler's first law and spatial analysis. *Annals of the Association of American Geographers*, 94(2), 284–289.

Mogel, L. and Bhagat, A. (2007). *An Atlas of Radical Cartography. Journal of Aesthetics and Protest*. Press Los Angeles, Los Angeles.

Monmonier, M. (1991). *How to Lie with Maps*. University of Chicago Press, Chicago.

Nüst, D. and Pebesma, E. (2021). Practical reproducibility in geography and geosciences. *Annals of the American Association of Geographers*, 111, 1300–1310.

Openshow, S. and Taylor, P.J. (1979). *A Million or So Correlation Coefficients, Three Experiments on the Modifiable Areal Unit Problem. Statistical Applications in the Spatial Science*. Pion, London.

Palsky, G. (1984). La naissance de la démocartographie. Analyse historique et sémiologique. *Espace Populations Sociétés*, 2, 25–34.

Palsky, G. (1999). Borges, Carrol et la carte au 1/1. *Cybergeo: European Journal of Geography* [Online]. Available at: https://journals.openedition.org/cybergeo/5233.

Pelletier, P. (2013). *Géographie et Anarchie. Reclus, Kropotkine, Metchnikoff*. Éditions du Monde Libertaire, Paris.

Peng, R.D. (2011). Reproducible research in computational science. *Science*, 334(6060), 1226–1227.

Philippe, R. (2014). Entre imaginaire et réalité : l'intention cartographique. *Revue*, 303(133), 8–15.

Poncet, P. (2011). Antigéographie. L'anticartogramme et ses interprétations. Report, WhereSciences.

Popelard, A. and Vannier, P. (2009). William Bunge, le géographe révolutionnaire de Détroit. *Le Monde Diplomatique* [Online]. Available at: https://blog.mondediplo.net/2009-12-29-William-Bunge-le-geographe-revolutionnaire-de.

Queneau, R. (1947). *Exercices de Style.* Gallimard, Paris.

R Core Team (2020). R: A language and environment for statistical computing. *R Foundation for Statistical Computing* [Online]. Available at: https://www.R-project.org/ [Accessed 13 April 2021].

Reclus, E. (1903). L'enseignement de la géographie. *Ciel et Terre*, 23, 29–38.

Robinson, A.C. (2018). Elements of viral cartography. *Cartography and Geographic Information Science*, 46(6), 293–310.

Schweikart, J.D. and Domnick, I. (2013). Thematic mapping between individual issue and mass media product. *Journal of Cartography and Geographic Information*, 63, 140–147.

Stewart, J.Q. (1942). A measure of the influence of a population at a distance. *Sociometry*, 5(1), 63–71.

Stienne, A. (2019). Lieux de pouvoir à Paris. Une carte qui dérange. *Le Monde Diplomatique*, 29.

Tiberghien, G. (2007). *Finis Terrae. Imaginaires et Imaginations Cartographiques*. Bayard, Paris.

Tobler, W.R. (1970). A computer movie simulating urban growth in the Detroit region. *Economic Geography*, 46, 234–240.

Tobler, W.R. (1979). A transformational view of cartography. *The American Cartographer*, 6(2), 101–106.

Zanin, C. and Lambert, N. (2012). La multireprésentation cartographique : exemple de l'atlas interactif des régions européennes. *Cartes & Géomatique*, 213, 39–64.

4

Mapping Flows and Movements

Françoise BAHOKEN
AME-SPLOTT, IFSTTAR, Géographie-cités,
Gustave-Eiffel University, Marne-la-Vallée, France

4.1. Introduction

From a semiotic point of view, the cartography of flows and movements may appear simple and unvaried, insofar as the modalities of representation are essentially based on the drawing of linear figures of varying widths that are eventually colored and oriented. However, interpretation of this type of map, which can represent material flows or movements (populations, goods, vehicles, etc.) or immaterial ones (currencies, ideas, etc.), is very responsive to the choices made by the cartographer at all stages of production. These choices influence the perception of the information, its comprehension and condition the communication of the transmitted message.

While several elements contribute to the understanding of a thematic map, this chapter focuses on a particular element: the flow map (or movement map) depicting spatial interactions by flows and movements. The flow data set, their specific analyses, their graphic transcription, their overall design and their meaning are all related to different kinds of challenges (theoretical and methodological, graphic, thematic, rhetorical, etc.). They must be explicit, discussed and put into perspective. In fact, all these components act on map figures, the perceived message and, ultimately, on its reading and reception by a given public. The perceived form of the sign is indeed fundamental in the communication of a flow map, as its

Cartographic Communication,
coordinated by Boris MERICSKAY.

misinterpretation can lead to a misunderstanding of the message, whether subconsciously or not (Bahoken and Lambert 2020).

Paying attention to the global process of flow mapping (from the structuring of data to the figurative elements used, as well as to the texts and colors chosen) is the essence of an approach that shows both geographical distribution of flows and movements, in particular, “their underlying structure” (Cauvin 1997), and their perceived pattern. That is probably why flow mapping appears normative in its implementation. In other words, if, as in any map, the final design matters in the cartographic communication process, the communication of a flow map is often more complex: it requires more attention and hindsight as a result of the intrinsic nature of the data.

The direct cartography of observed flows, without prior thought, is in fact generally not applicable due to the manifestation of various graphic effects that must be reduced or even removed in order to produce a readable and thematically interesting image. We first mention the “spaghetti effect”[1] (Bahoken 2014a, 2014b), which involves filtering the flow information or the features to draw, reducing their resolution by aggregating flow data, or the “amalgam effect”, which implies breaking down or transforming the data (Bahoken 2016a). “The spaghetti effect” is the more important of the two. It reflects a graphical complexity[2] in the perception of the depicted flows. Other effects can also be added, such as the “alignment effect”, the “route effect” and the “join effect”, in the case of flow maps made at the global level (we will return to this).

The dual function of information and communication of the flow map thus appears to be more difficult to fulfill than for other thematic maps. Attention must then be paid to the different pillars of the cartographic process (MacEachren 1995); in other words, the processing of numeric and geographic data on the one hand and their articulation to build a significant flow map on the other, including their semiotic enrichment. For a given subject (demography, economy, etc.), what will therefore change from one flow map to another has more to do with the underlying theoretical and conceptual framework, that is, the way in which geographical space is considered in the cartographic process, the methods implemented and the tools used. The objective of this chapter is both to propose a summary of the elements

1 This effect becomes apparent as soon as the number of locations exceeds 10 (for 10 locations, 10×9 means 90 links to represent on the map).

2 The effectiveness of a map in conveying a message is evaluated using three measures: (1) visual complexity, corresponding to the effect felt at the time of perception ; (2) color complexity related to the style and hues (single, gradient, combined) and (3) graphic complexity related to the quantity of figures represented.

involved in the construction of flow and movement maps and to review the associated communication challenges in light of the technical developments in recent years.

The theoretical and methodological framework of flow mapping are introduced first, and then their implementation, which leads to the main types of flow maps, is proposed in light of recent technical developments.

The geographical challenges of the representation of flows and movements are explained and illustrated in a third step through an exercise of cartographic multi-representation of the globalization of containerized goods flows.

4.2. Theoretical and methodological considerations of the cartographic construction of flows

Theoretical and methodological aspects are essential in the elaboration of a flow map, and the cartographer must be familiar with them and, at the very least, question them. The type of map that we can make depends above all on the properties of the flow data materialized in an origin-destination (OD) matrix.

In the same way, the interpretation of the map, in terms of a network of relations, exchanges, flows or even movements, depends on the characteristics of the matrix. It involves the communication of cartographic information in the form of a flow line.

4.2.1. *The flow line*

The graphic representation of a flow (F) expresses the transfer of a measured quantity (of people, goods, values, etc.) between a pair of places of origin (i) and destination (j). The object of study is a bi-point (i, j) that mobilizes two geometries: the points (XY) associated with the coordinates of the ODs forming the pair of locations, i(x_i, y_i) for the origin and j(x_j, y_j) for the destination, and the line expressing at least the existence of a relationship between (i) and (j).

The line is a perceived form of displacement. It will either conceptually structure the map as "image variables" (Bertin 1967), or it will be based on a bi-point and be sensitive to a variation of its morphology. Whatever its geometry, the semantics of the line can also be enhanced by the application of semiotic signs, such as shape, size or color (Figure 4.1).

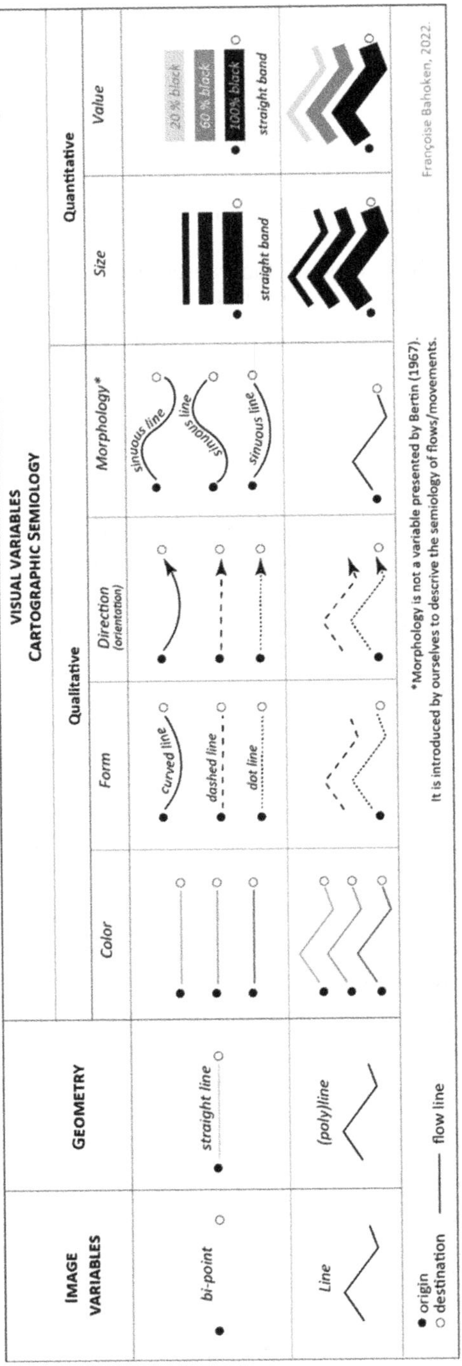

Figure 4.1. *Geometry and graphic semiology of the flow line. For a color version of this figure, see www.iste.co.uk/mericskay/cartographic.zip*

Beyond the shape of the line (straight, curved, dashed, etc.), its morphology is important because it allows us to distinguish between the representation of flows based on a bi-point and that based on a path in space. These two types of spatial implementation of flows correspond to two concepts: the OD flow one, which is represented in the form of a straight or curved line, to connect two points (and whose morphology may vary, such as for esthetic reasons), and the flow affected on a network, whose layout (and therefore morphology) on the map does not vary. In both cases, these perceived lines are associated with points that also structure the image.

4.2.2. *The Links-Nodes logic of cartographic reasoning*

The duality of the representation of flows mobilizing a line and at least two points translates the duality of a theoretical reasoning of the flow map, proposed after Tobler by Grasland (2009), articulating the representation of spatial interactions (by lines) and that of positions (by points). This posture corresponds to a Links-Nodes logic of reasoning, which calls upon all the aspects of the available data, in other words, the complete matrix. The Links-Nodes cartographic reasoning finds its origin in the structure of the starting table, that is, the cross-table (an OD matrix) where by convention, the rows show the entities of origin places and the columns show those of the destination. The cell defined at the intersection of a row and a column describes the existence, or absence, of a relationship between a couple of places, or even a transferred quantity.

The core part of the matrix corresponds to the interactions represented by a line. They link places, represented by points (or areal features), corresponding to the margins of the matrix, which initially express the total emitted (or received) by a place located in line (or in a column). Indicators can then be computed on these OD places and on the links, such as the volume of flows involved with each of the places and that exchanged between them (we will return to this).

Figure 4.2 illustrates the cartographic representation of a flow matrix of population, especially migrants, as the so-called stocks (discrete quantitative characters). This stock of individuals can be translated into larger or smaller arrows to illustrate migrations and/or proportional circles in order to show the effect of migrations on the distributions of departure and/or arrival places.

Attention is drawn to the fact that the location points reflect the information available on the margins of the matrix, in this case a volume of flows, and not the information carried on the possible diagonal of the matrix. Using the diagonal is a special case that leads to the mapping of inter- and intrazonal information[3]. In the same way that the choice of a line/arrow/dot depends on the part of the matrix used (core/margin), the choice of an arrow/line to materialize the displacement depends on the format of the data located in the core of the matrix and thus on the type of OD matrix.

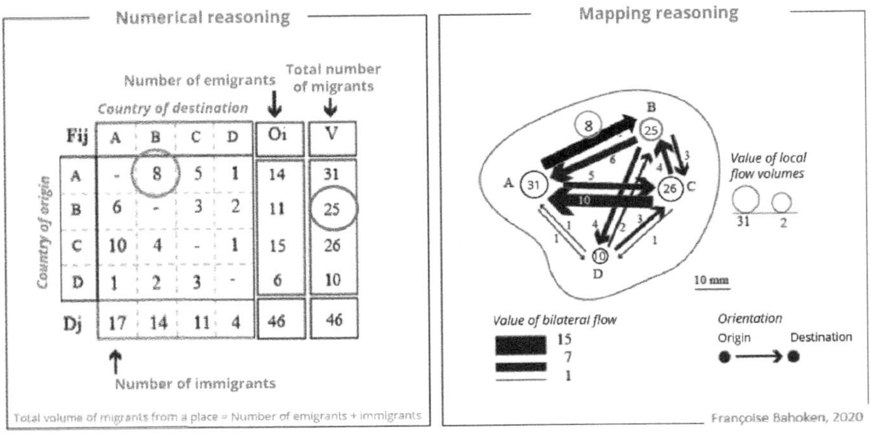

Figure 4.2. *Links-Nodes reasoning and mapping (flows, locations) (Bahoken 2020). For a color version of this figure, see www.iste.co.uk/mericskay/cartographic.zip*

4.2.3. *From the type of matrix to the type of mapped flows*

Three main types of matrices can be identified according to the general property of symmetry of the values located on either side of their main diagonal line. Their asymmetrical, symmetrical or anti-symmetrical nature refers to three main forms of flow maps, which themselves involve three combinations of line/arrow features (Figure 4.3). These three types of matrices are based on two theoretical registers: graph theory (for an analysis of data morphology) and gravity logic (for an analysis of spatial or territorial interactions).

3 The information on the diagonal is intra-zonal flows, and does not formally correspond to a flow defined as the transfer of individuals between ODs. The diagonal corresponds to individuals who remained "stationary" at the observation level, or whose movements did not cross the area limit.

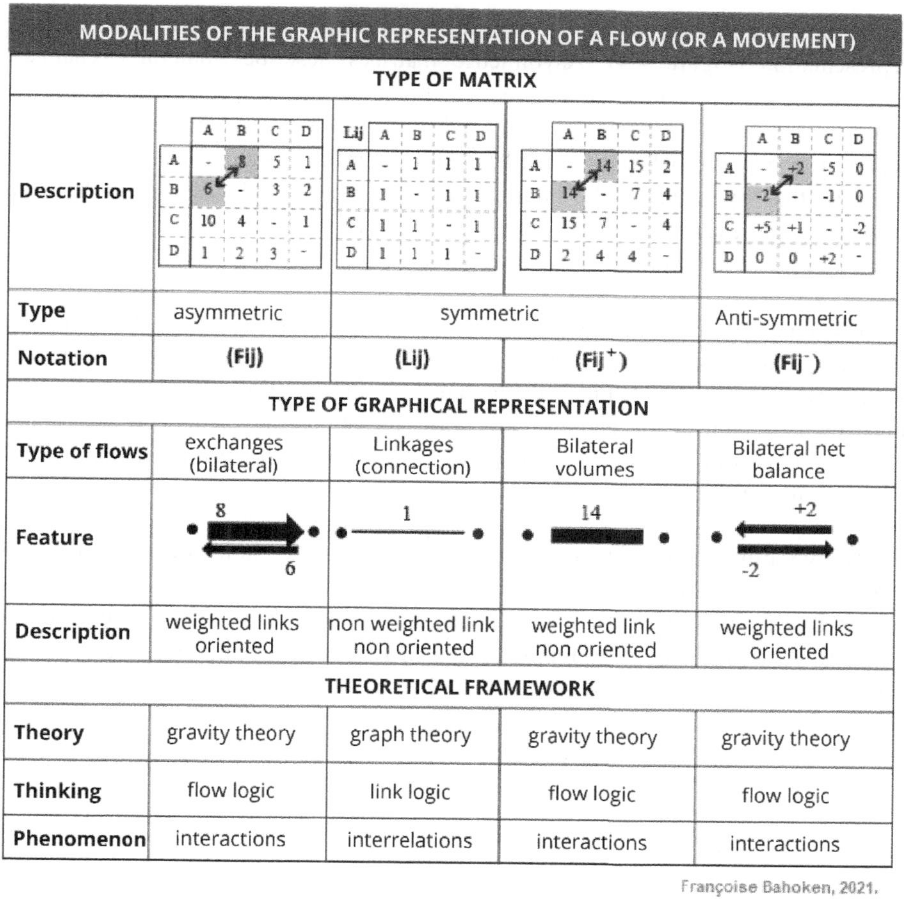

Figure 4.3. *Matrixes and flow map types (Bahoken 2021)*

"Symmetric matrices" have identical values on both sides of the main diagonal line. Therefore, they must be mapped as a non-oriented line; this line can be weighted and take the form of a band.

"Non-symmetric matrices" have different values on either side of the main diagonal line; their representation requires the use of at least one arrow (oriented band). Among these, the "anti-symmetric matrices" have identical numerical values on both sides of the main diagonal line but opposite signs; their graphic translation requires a single arrow. The "asymmetric matrices", whose values on both sides of the diagonal line differ, necessarily require two arrows of different widths and in opposite directions.

These three main types of matrices thus correspond to three main types of flow representation (Figure 4.4). Asymmetric matrices describe bilateral exchanges. Symmetric matrices map a volume of bilateral flows[4], thus describing the "foundation of the trading system" (Tobler 1982). Anti-symmetric matrices represent a bilateral balance (or net balance), describing bilateral net transfers that illustrate the "dynamics of the trading system" (Tobler 1982). It is possible to go further by representing, for example, a combination of volume/balance maps to reflect the asymmetry of flows[5] (Bahoken 2016a).

The characteristic net balance[6] of an anti-symmetric matrix is normally the subject of two mirror maps: one for positive values and the other for negative values, with figures of identical widths but different directions. Particular attention must then be paid to the choice of the flow direction (i,j) versus (j,i), especially when the places in rows and columns diverge, such as those describing home-work flows or imports-exports, for example.

To avoid this mode of representation that leads to the choice of one of the two maps, it is possible to produce a single map by merging these two net balance maps to generate a single map (Bahoken 2016a). The operation consists of merging the two arrows, (i,j) and (j,i), by removing the two arrowheads and applying a hue in colored opposition[7] in order to evoke the direction of travel (Figure 4.5).

4 The bilateral flow volume is the arithmetic sum of the values of the links on both sides of the main diagonal line, of a non-symmetric matrix: (Fij) + (Fji).

5 The flow asymmetry is the ratio of the balance to the volume of flows, computed over locations and over links. The resulting representation of the links either needs to be interpreted as a variation of the balance map or as the volume map (since the result of the balance/volume ratio cannot be used to size a sign). Since the flow asymmetry cartography in Figure 4.4 is understood as a variation of the volume map, it combines the volume map, at the places level, by a proportional circle with a discretized color by the flow asymmetry value and, at the link level, the bilateral volume map by a band. The flow asymmetry value is the ratio of the balance to the volume calculated on the margins.

6 The net balance of flows is the difference between the values of incoming (D_j) and outgoing (O_i) flows; the bilateral balance of flows (or net transfer) is the difference between the values of the links on either side of the main diagonal line of a non-symmetric matrix.

7 The use of the colored opposition principle to represent flows is therefore only applicable for anti-symmetric matrixes; otherwise, it leads to obscuring part of the information.

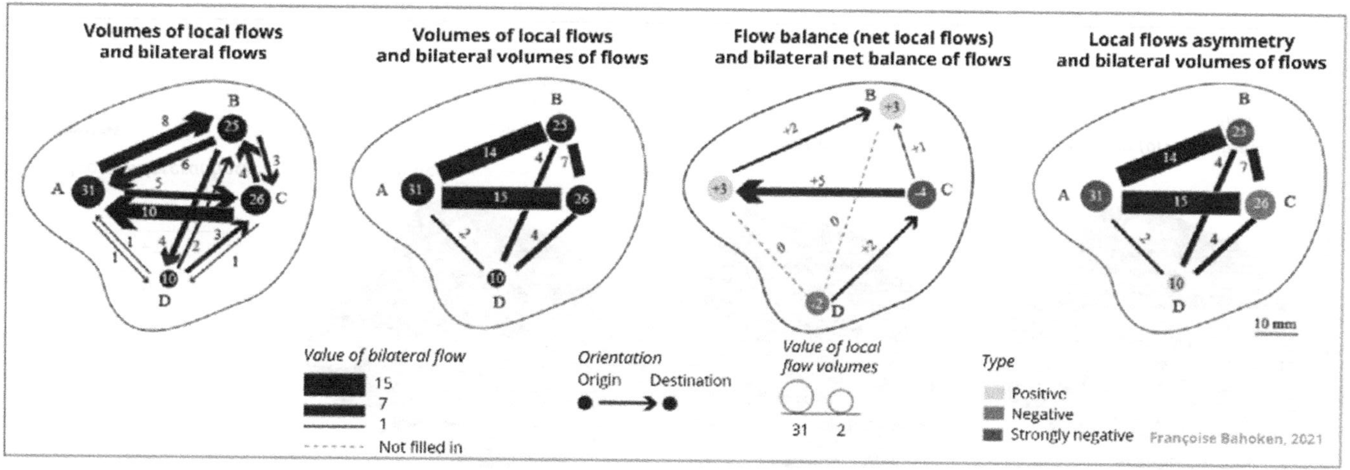

Figure 4.4. *Main types of flow maps (Bahoken 2021). For a color version of this figure, see www.iste.co.uk/mericskay/cartographic.zip*

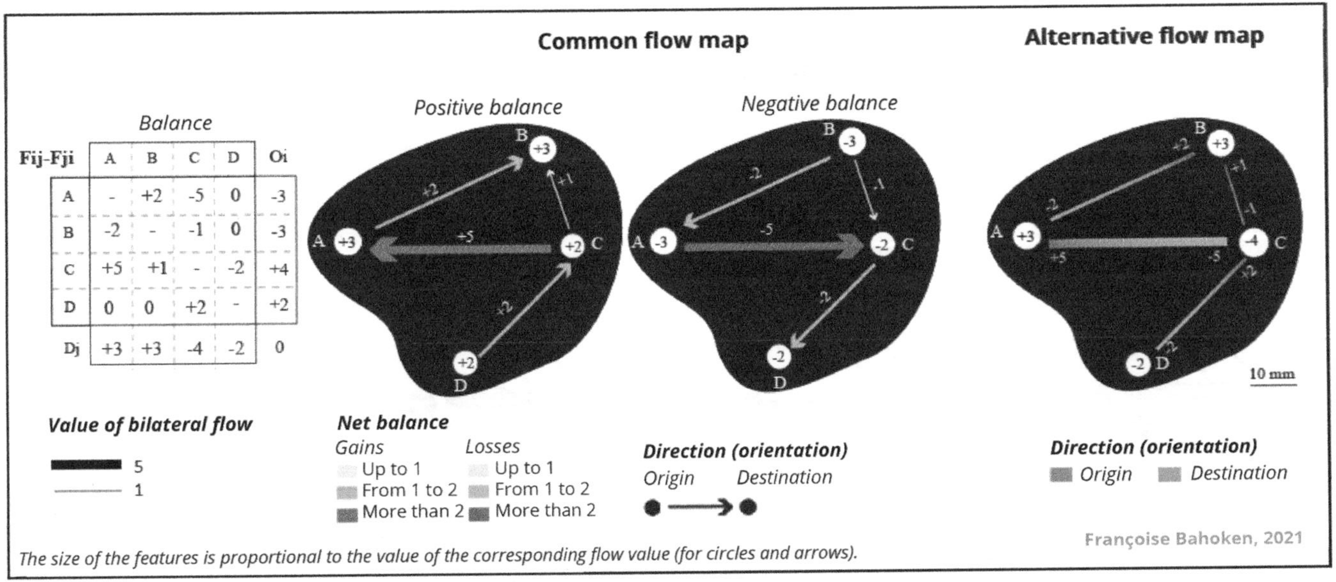

Figure 4.5. *Variations in the representation of bilateral balances (Bahoken 2021). For a color version of this figure, see www.iste.co.uk/mericskay/cartographic.zip*

This type of bilateral balance representation has several advantages, such as summarizing information in a single map while avoiding the traditional arrow and its graphical constraints. The representation of the flow dynamics in colored opposition also helps in the interpretation of the map: it can be read from the perspective of origins and/or destinations. In addition, this type of map also allows for the original representation of the poles of emission and attraction of flows.

Although the single or double sign (line/arrow) depends on the type of matrices, its morphology also reflects a flow/movement map type.

4.2.4. *From flow to movement*

The drawing of the flow line/arrow is founded on a point (a bi-point) or linear spatial layouts, whose significance depends on knowledge of the mode of transport between OD places. The ideal case for a complete flow representation is to know the exact path taken by the flow, but this is quite rare. The information generally available is that of ODs, which can only define an approximate linear form, imperfectly describing a movement; as a result, information about the mode of transport, as well as the routes taken, is incomplete or even non-existent.

Whether the line/arrow geometry is sinuous or not, representing transport movement is not without consequences to the interpretation of the map, as the norm of the underlying vector reflects the "map distance" perceived by the reader (Bahoken 2016b).

Three forms of distances are effectively used in the cartographic representation of flows:

– the "cartographic distance", which is perceived on the map by the length of the link and which is interpreted in terms of flow or movement, depending on the morphology of the line;

– the "geographical distance", which is geometric and generally approximated by the Euclidean distance (the flight of a bird gives the link its rectilinear aspect);

– the "empirical (or measured) distance", related to the generalized cost of transport, whose cartographic transcription follows the transport infrastructure used or the path in space.

One of the key elements in the process of flow mapping is therefore the knowledge, or lack of knowledge, of the transport infrastructure, which gives the line its sinuous shape, approximating this network with varying degrees of accuracy. This being the case, four main approaches can be used to map a movement:

– The "geo-mapping approach" consists of acting on geographic information (the territorial mesh) in order to choose the space in which to define the flows to be represented, its resolution, its scale and even its spatial implementation mode.

– The "mathematical approach" is based on a transformation of the information to be represented, for example, by weighting it, modeling it or filtering it using spatial variables.

– The "(carto)graphic approach" is founded on the drawing of flows on a plane that can be provided with a cartographic projection, as well as on its semantic enrichment with the help of visual variables. It is essentially based on graphic principles, possibly adapted to the map as a modality of graph with strong constraints of positioning in an often-geographical space.

– The "infographic approach" is linked to the screen medium; it consists of acting on the temporality of the representation in order to show the dynamics of a flow or a movement in an animated or dynamic way using adapted visual variables such as flickering, animated particles or droplets (Côme et al. 2021).

Figure 4.6 illustrates the importance of the type of flow/movement concept and their geometries, which lead to the qualification of the two types of maps (flows/affected flows on network) as a graphic translation of a real or approximated transport movement.

A linear implementation means that the medium of the map (the transport infrastructure) is the flow information itself, which leads to the representation of assigned flows on a network. Such maps are characteristic of richer flow information since the mode of transport, the quantities in movement and the sequence of transit locations are known and allow the trajectory to be represented spatially, making the map coherent with the geographical reality of the studied space.

In the case of a bi-punctual implementation, the bird's-eye view illustrates the transfer from one place to another, independent of spatial constraints, as if geography had no influence on the movement. The approximate shape of the line drawn between ODs generates an inevitably partial map, reflecting the lack of information about the exact path taken by the flow. The difference between "flow" and "movement" maps therefore lies in the consideration of spatial constraints on movement in the cartographic construction of the flow (Bahoken 2016b; Bahoken et al. 2016) and not just in their mathematical modeling that is performed upstream of the representation step or in the graphic design occurring downstream.

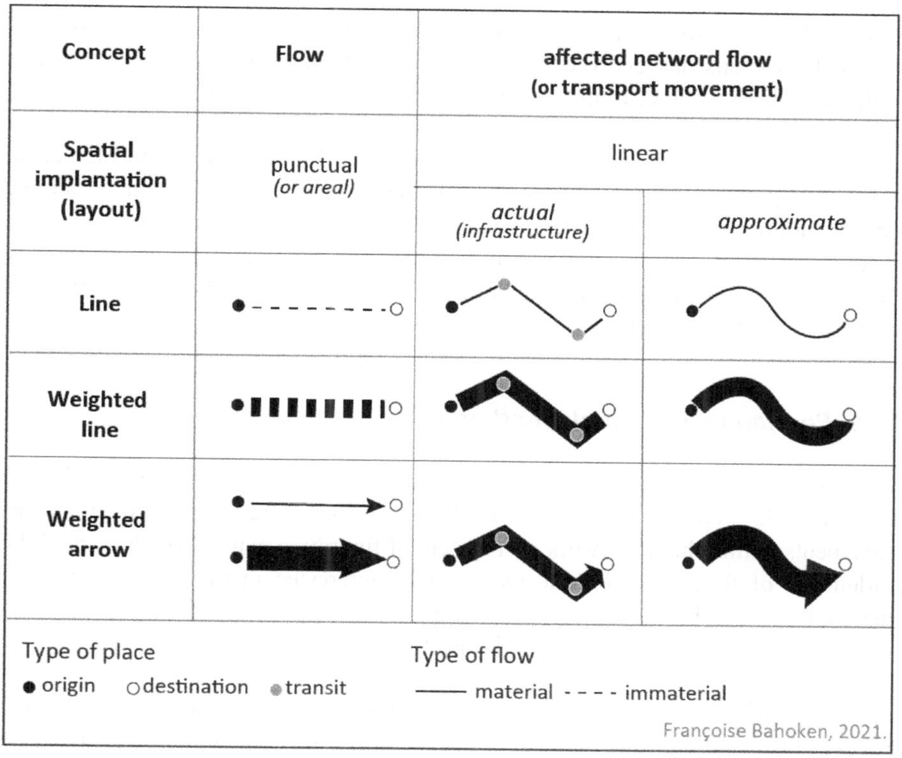

Figure 4.6. *Geometry of transport flows and movements (Bahoken 2021)*

Not all these aspects related to cartographic design are trivial, and the possibilities of their shaping are not extensible. They also depend on the tool, the choice of which will have consequences on the type of flow maps to be represented and on their style.

4.3. The cartographic image of the flow: between continuity and technological disruptions

The cartographic representation of flows, its underlying methods and its usual principles are part of a long history of thematic cartography. It begins with the representation of "transport movements", an expression borrowed from the French engineer Minard, who had the idea of visually representing statistics on movements and, as a result, produced earlier maps of flows/movements. His work thus lays the foundations of a method that is still relevant today.

From Minard's first maps to the interactive geovisualizations of the Geoweb, the face of flow maps has evolved with theoretical and methodological advances, in conjunction with the development of dedicated tools (Bahoken 2019).

These changes, which involve the processing of information (both mathematical and geographical), contrast sharply with their mapping based on a simple representation method. The representation of flows, carried out according to a "flow logic", consists of acting on the width of the line between ODs to graphically translate the significance of the transferred flow, possibly on the sizes of the OD locations, in order to illustrate the effect of these flows on their spatial distribution.

4.3.1. *Permanence of an old method of representation*

As a precursor in this field, Minard devoted his retirement as superintendent of the École de Ponts et Chaussées in Paris to the representation of transport movements within the framework of a system of figurative maps[8], which marked the golden age of flow mapping. He was initially interested in the representation of "material flows" (based on an infrastructure) before generalizing the process to "immaterial flows", unaffected by a transport network.

Minard, whose anthology is available in Grootens and Rendgen (2020), uses a process of distributive flow mapping, popularized as the Sankey diagram[9]. He used the visual variable dimension (which will become size) to map transport movements at a time when it was mainly used for populations. He applied it to different point locations to characterize flows according to a geographic "location logic" (receiving cities, shipping emission ports, etc.), before implementing it online. For example, he represents road traffic by characterizing the sections of the infrastructure by a variation in their width[10].

8 A figurative map is one that does not exactly respect the geography of the concerned spaces, in order to focus, by their representation, on the perception of the flow information to the detriment of their geography.

9 The Sankey diagram is named after the British mechanical engineer who was successful in using this flow distribution process to represent the energy flow of an engine. The process takes the form of a system of colored bands of varying sizes and morphologies, drawn between OD points that are located on parallel vertical axes.

10 The cartographic representation of the use of a road network known as a "load line graph" had previously been introduced by Harness in 1836 in order to count the flow of vehicles to/from Dublin.

Minard also uses color in different ways. His first map, which was complete in black and white, presents several levels of depth that touch upon the use of the visual variable value in order to distinguish different more or less complete routes (Palsky 1996). He uses color in its qualitative sense to illustrate the movement of wines and spirits; for example, the colors identify the various transportation routes (hydrographic, sea, land) or the direction of movement (export versus import).

If Minard's cartographic work seems purely descriptive, it is because it is a precursor. Indeed, it was not until the end of the 19th century and the publication of Ravenstein's empirical laws of migration (1885, 1889), accompanied by maps, that a new era of flow/movement cartography began (Bahoken 2019), linked to the practice of graphic statistics that would continue to develop. The development of statistics, which was related to the advent of demography, would indeed lead to the progressive implementation of a theoretical and methodological corpus that was specific to the analysis of spatial interactions[11]. This corpus, often accompanied by maps, was enriched over time by conceptual transfers from the hard sciences, particularly physics (which led to the gravity analogy). More innovative methods of representation, allowing the results of these different theoretical advances to be represented, would then progressively develop from Ravenstein's map called "Currents of migrations" (1885) and those proposed by Thornwaithe (1934) based on physical principles. This position is also that of Tobler (1995), who took this up in the 1970s, in particular to demonstrate Ravenstein's empirical laws and to develop several innovative cartographic representations of flows/movements.

4.3.2. *From a discrete approach to a continuous flow/movement approach*

Tobler, a geographer and cartographer whose approaches combine mathematics and computer science (Cauvin 2020), played a major role in the development of a digital and cartographic approach to flows and movements. He was interested in their modeling and representation (Tobler 1970b) in terms of processes, including the tools to implement them (Tobler 1970b, 1979b).

Tobler (1979a) first develops a cartographic method associated with a model, making it possible to represent attractions (flows according to a logic of places) before focusing on interactions, such as flows/movements between places. The first problem he seeks to solve is then cartographic. It concerns the representation of movements, taking into account the geographical positions of the pairs of ODs and the values of the flows that connect them. Tobler then explores the construction of a

11 See in particular the thesis of Poulain (1981), which presents a history of it.

simple and operational model, allowing him to discover movements and places of attraction. He makes extensive use of cartography to illustrate the obtained results in parameterizing his model: estimate flow values from the distances between ODs and the marginal sums of the matrix, which corresponds to an inversion of the gravity model. His work leads him to use matrix calculation, which he thus transfers to the analysis of social and spatial interactions.

Tobler (1979a, 1982) shows that the best way to map flows is to represent vector fields leading to attractions (Figure 4.7 on the right), calculated on what he calls "the negative component of the flows", in other words, on an anti-symmetric matrix.

His overall thinking (Tobler 1987) led him to lay the foundations for a cartographic representation of OD matrices, establishing three types of flow maps: discrete, continuous and animated maps[12], which are differentiated "according to the way they show movement" (Tobler 1970a). Figure 4.7 shows these two paradigms: discrete (left) and continuous (right), as conceptualized by Tobler (1981).

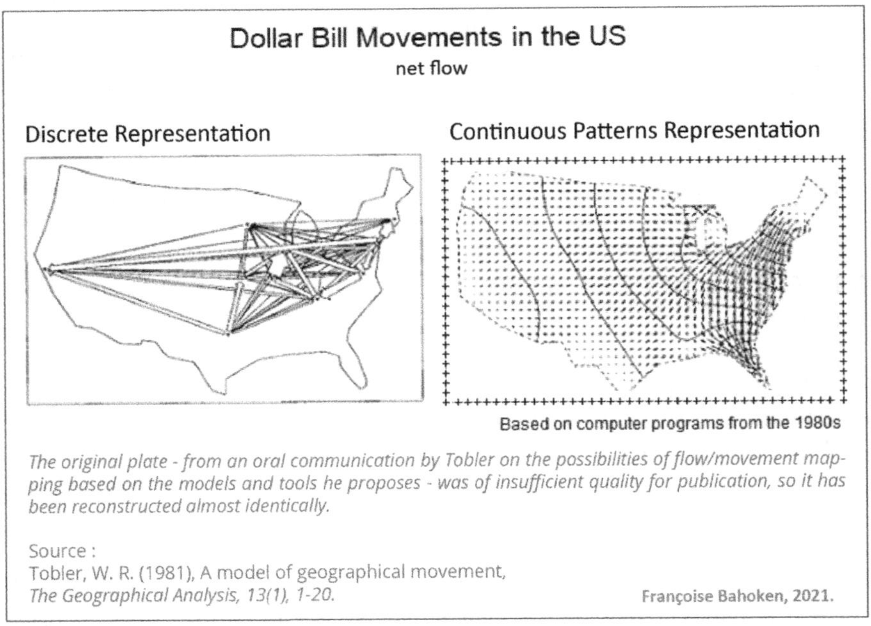

Figure 4.7. *Illustration of discrete and continuous flow/movement cartography (source: Tobler 1981)*

12 Animated maps are no longer a flow/movement map model as such, as movements can now be presented on static or interactive/animated maps.

4.3.3. *Mapping flows in the computerization era*

The evolution of flow iconography is undeniably linked to the tools allowing its production. Although, as Joliveau (2011) notes, the advent of Geoweb[13] has changed the use of online maps and their production, their characteristic geodigital medium nonetheless remains a new "vanishing line" (Plantin 2012). This means that the contemporary map is part of this long history of cartographic support, from the first clay tablets to the historical maps of Mayan travel, passing through the Peutinger Table to take the contemporary form of current maps.

Although still in tune with its era, if the flow map has been able to evolve with theoretical and methodological advances (Bahoken 2019), it is above all thanks to the development of dedicated tools. Flow, which has always been the poor relation of thematic cartography, has always required specific tools developed by the researchers themselves. This historical restriction of flow tool development to the academic world has not prevented their evolution or indeed their regular adaptation to the languages of the moment. Tools allowing the drawing of flows have now become essential, thanks to the current possibilities of deployment in different environments, in relation to the availability of big data sets that are often free and open, giving rise to increased needs for visualization.

The computerization of the 1970s, followed by the development of GIS up until the interactivity of web maps in 2010, has indeed profoundly renewed, as well as facilitated the cartography of OD matrices, especially those that are dense. Unlike traditional maps, especially choropleths, the tooling of flow maps has usually been the work of the cartographers themselves in response to the specific representation needs of this type of data. Although there are no generic tools (we will come back to this), flow computerization will nevertheless modify the way in which maps are rendered, the way in which they are produced, and certain previous technical innovations.

Several technical problems and obstacles have been overcome with computerization and new algorithmic possibilities. The geovisualization of OD matrices also benefits from the general craze for relational data visualization. Old methods of representing flows are thus being revived and gradually made accessible for all. This is particularly the case for the graphing of routes, including for large matrices, which is now possible using a technique known as edge bundling. This purely graphical method has been developed with the objective of reducing the graphical complexity of the flow map by merging the lines to approximate the

13 For more information on the Geoweb and its impact on cartographic communication, see Chapters 5 and 6.

representation of the main routes. Several proposals have been made regarding this, including those of Phan et al. (2005), Holten (2006) and Melançon (2010). On the one hand, a specific effort has been made in the automation of spatialized graph design from a perspective that directly involves the representation of flows (Jenny et al. 2018). On the other hand, there is a renewal of graph design that enriches the palette of graphical possibilities for flow/movement lines.

However, these advances have not been made without difficulty. The cartographic representation of flows has always been somewhat neglected. It also had never truly been implemented in CAD and GIS tools until recently. A first explanation is that flow maps require considerable input data processing. A second explanation is that their construction is generally not taught in the French university system; at most, they are the subject of one or even two tutorial sessions in some specialized courses. Flow mapping is not the subject of dedicated manuals either, which does not facilitate its appropriation. This is why their construction has always been the subject of dedicated instrumentation, resulting from the development of specific programs and tools by people who are specialists or interested in this matter.

Since the initial proposal of the automatic origin-destination flow data mapping program Flowmapper (Tobler 1979b), which was developed in Fortran and is currently being ported to the R ecosystem (Lambert et al. 2021) or JavaScript Observable, the tool has gone through several versions. First integrated into stand-alone tools, it was then implemented in various computer-assisted mapping environments (Philcarto, Cartes & données, Magrit) and GIS environments (ArcGIS, MAPINFO, QGIS). The 2010s saw the advent of mapping applications such as Arabesque (Côme et al. 2019), dedicated to flow mapping developed in Java (Boyandin 2010; Boyandin et al 2011), and then in JavaScript. Mapping flows today means dealing with several families of tools and plugins with different functionalities and performances that respond to the specific needs of distinct communities: business cartographers, data journalists, developers in computer graphics or scientists using languages such as R or Python.

4.3.4. *Cartographic image of the flow on the Geoweb*

Contemporary flow maps, such as cartographic production on the web, have a strong visual impact supported by modern stylization. They are largely overvalued in their appearance because of the "innovative" and "modern" character of the medium. Geoweb flow maps also benefit from a new form of authority that is specific to information and communication technologies (Bahoken et al. 2020). This authority is related to their "performative potential, that is to say, their capacity to induce interaction with the one who looks at the image or work" (Amilhat Szary

2018). The appearance and style of the image are fundamental in the reception of the cartographic message, and the Geoweb responds well to this: the digital medium, due to the almost infinite range of possibilities for stylizing maps, allows for a renewed appearance of flow maps. Figure 4.8 shows four flow maps based on the same data set developed with the online application Arabesque (Côme et al. 2019). With this type of tool, in just a few clicks and from a web browser, the cartographer can filter or aggregate the data, customize the layout, play with the style of the mapped flows, and change the map backgrounds.

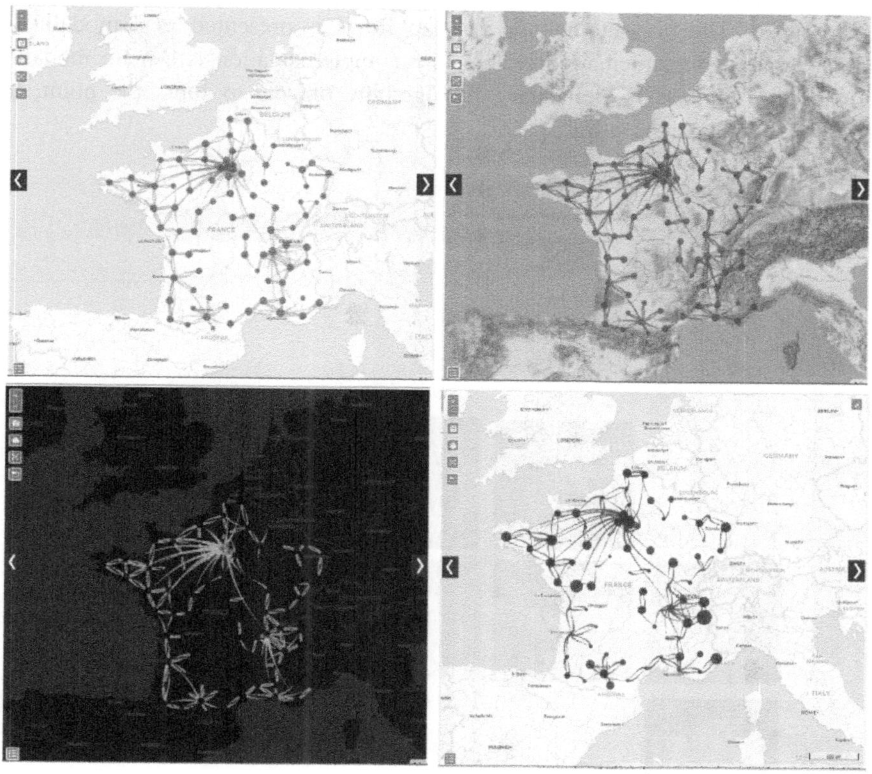

Figure 4.8. *Example of flow map stylization with the Arabesque application (Bahoken 2021). For a color version of this figure, see www.iste.co.uk/mericskay/cartographic.zip*

The communication of flow information by cartographic means, including geo-numeric aspects, should not, however, leave aside the graphic semiology that allows maps to be meaningful[14]. It is indeed important to recall the importance of the cartographic language in the transmission of information because of the role played by the application of retinal variables. Of the three principal modes of spatial implementation (point, line and surface), the flow map requires two specific modes: the bi-point and the line, which have already been mentioned. Figure 4.9 is inspired by a proposed presentation of elements of cartographic representation on Geoweb (Mericskay 2016) to specify the case of flows and movements, their spatial implementations and their geometries. It introduces the possibility of a bi-point implementation, so that OD flows and those assigned to the network are differentiated. The bi-point method of map flows is presented as a modality of punctual spatial implementation, where linear implementation is a main modality, bringing the number of elementary (geo)graphic objects to four (the point, the bi-point, the line and the surface).

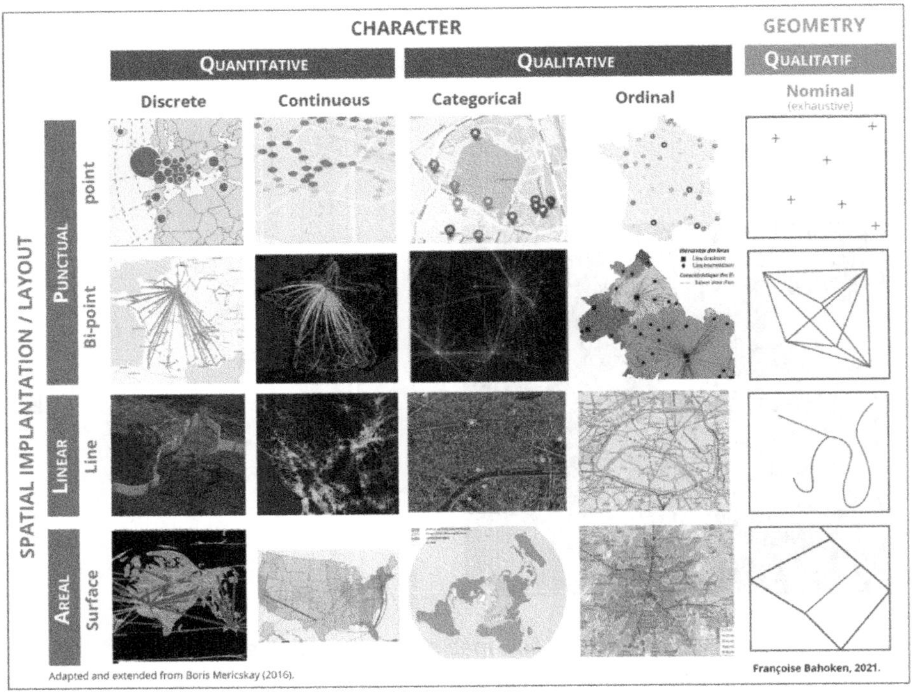

Figure 4.9. *The image of the flows and networks communicated in the Geoweb (Bahoken 2021). For a color version of this figure, see www.iste.co.uk/mericskay/cartographic.zip*

14 See Chapter 6, which revisits the graphic semiology of the Geoweb.

Several formatting configurations thus appear, depending on the nature of the represented flow data, each of which conveys information differently:

– the representation of a continuous quantitative characteristic leads to the drawing of a linear figure, then shading its color by applying a discretization, carried out on a continuous variable;

– the representation of a qualitative categorical characteristic consists of coloring the linear figures according to the qualitative class to which it belongs;

– the representation of a qualitative ordinal character consists of representing a distribution of ordered flow values by adapting the semiology of the vertices in such a way as to translate the hierarchy of places resulting from an analysis, such as for dominant flows (Nyusten and Dacey 1961).

Beyond these aspects related to data, as well as to the cartographic rendering obtained with one or another of the contemporary tools, flow mapping is subject to specific challenges that are all the more difficult when expressed at the global level. The geographical constraints exercised at this scale place the geographical space at the heart of the flow mapping challenges that are related to the specific cartographic communication of flow/movement information.

4.4. Challenges related to flow mapping through the example of global maritime flows

To examine and discuss cartographic communication challenges related to the representation of flows, we propose an exercise of multi-representation[15] around the example of international trade flows, observed at the global level. To synthesize a series of challenges related to the intention of the map[16], we use an asymmetric matrix that links six economic macroregions by international containerized flows observed in 2006.

These flows correspond to the number of containers transported from one region to another, expressed in thousands of twenty-foot equivalent units (TEUs). We start from a representation of these flows that were initially published in French in the *Atlas de l'Europe dans le monde* (Didelon et al. 2008), which we consider a reference image for this type of map (Figure 4.10). The objective here is to show the

15 This multi-representation challenge is developed in Chapter 1. It is also implemented in Chapter 3, around a world population data set.

16 Challenges relevant to the reception of the map are not presented. See an example in (Bahoken et al. 2020).

consequences of different theoretical and methodological choices on the associated maps. The initial map will thus be deconstructed/reconstructed from different perspectives (semiotic, projective, background map, etc.), to allow us to ultimately propose and discuss a series of (carto)graphic representations based on the same data while conveying the message in different ways.

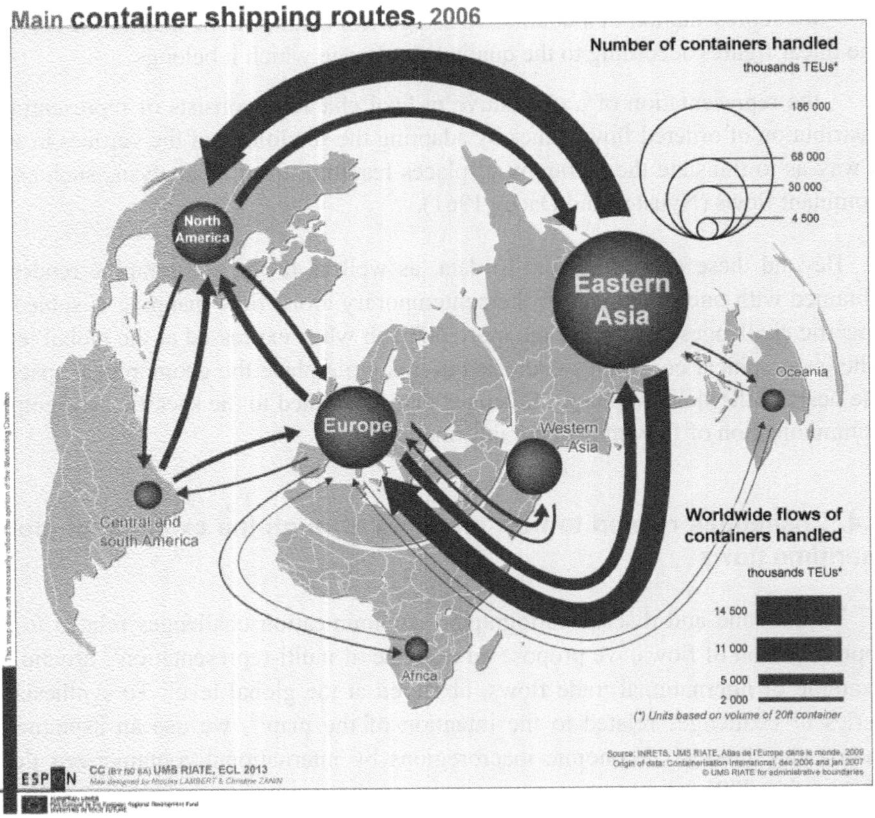

Figure 4.10. *Main maritime routes for container transport. For a color version of this figure, see www.iste.co.uk/mericskay/cartographic.zip*

4.4.1. *Challenges of flow mapping at the global level*

The globalization map is subject to specificities at the global level when the latter is apprehended as a territory (Didelon 2013). This position implies considering the particularities of this scale, which particularly result from the sphericity of the earth and the alternation of land and sea, which generate graphic effects or constraints in terms of their mapping in the form of flows. The informative character of the map and its coherence requires that its production be part of a double register. First, the "thematic" register is the subject of the map. For example, the representation of global maritime trade flows differs from that of international migratory flows. Just as these fundamental themes are different, mapping maritime trade flows should not be identical at migratory demographic flows. Second, the cartographer's "intention" is to communicate one piece of information rather than another in response to a social demand that is relevant to the reception of the map. This social demand for cartographic information refers to the "destination" of the document, as well as to its "reception" by a public at a given time, in a particular geopolitical or socioeconomic context. Aside from the choice of the cartographic projection system, the challenges of a cartographic representation of flows at the global level vary. On the one hand, they are primarily linked to the sphericity of the Earth; indeed, the transition from a sphere to a plane induces the choice of a projection that is adapted to the type of flow due to a "joining effect". This effect corresponds to the fact that certain projections lead to the interruption of flow lines that are expressed at a very long distance, thus improperly suggesting an interruption in displacement. These challenges are associated with the alternation of land and sea, which generates two particular constraints. The first is an "alignment effect" (or position effect) (horizontal, vertical or oblique) of the places of interest, which hinders the clear perception of the flows because of too many overlapping signs. The second constraint is a "route effect" that implies taking into account transportation in the mapping process of land and sea flows at the risk of thematic incoherence (in the specific case of maritime flows).

4.4.1.1. *The alignment effect and the curvature of the lines*

The "alignment effect" is the result of a worldview based on the traditional center-periphery model that generally focuses attention on a few so-called central places. These places structure macro-region areas, within which other places are under their influence. The polarization of the world economy by these central places, mostly located in the northern hemisphere, reduces the map to a defined space of interest that is limited to a half map of the world, potentially biased by the horizontal alignment of these centralities (Figure 4.11).

Consequently, this positional effect prevents the visualization of the main spatial patterns, such as those of the Triad, because of the overlapping bilateral flows that are visible in the high latitude regions with large exchanges. One solution to this problem is to change the shape of the rectilinear arrows and turn them into curves (Figure 4.12). The curve is visually effective in that the map is now clearer due to the reduction of two complexities: visual perception and understanding of information. The map is more readable and therefore more pleasant to look at, thus changing its effectiveness. However, the improvements brought about by this arrow design do have possible consequences on the perception of information. Overuse potentially leads to the misrepresentation of information because of the role played by the cartographic distance perceived by the length of the line (it is more important for curved lines than straight ones) in the interpretation of the map. In addition, the curvature of the stripes/arrows increases their perceived visual area, suggesting that the information would have changed between Figures 4.11 and 4.12.

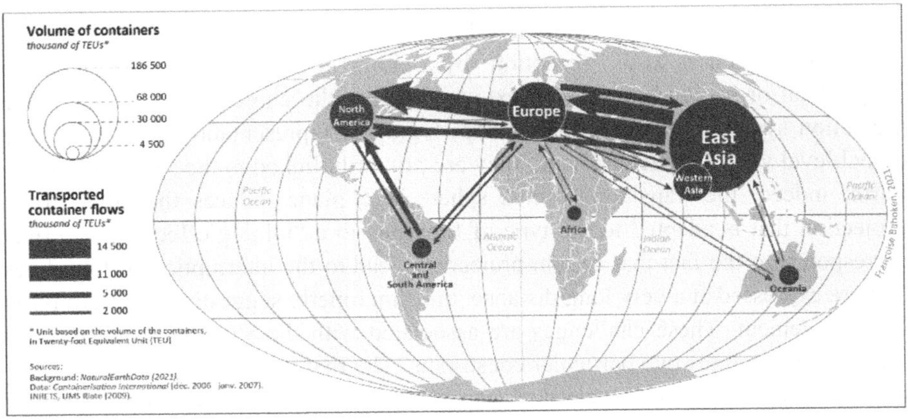

Figure 4.11. *The alignment effect of global maritime flows (Bahoken 2021). For a color version of this figure, see www.iste.co.uk/mericskay/cartographic.zip*

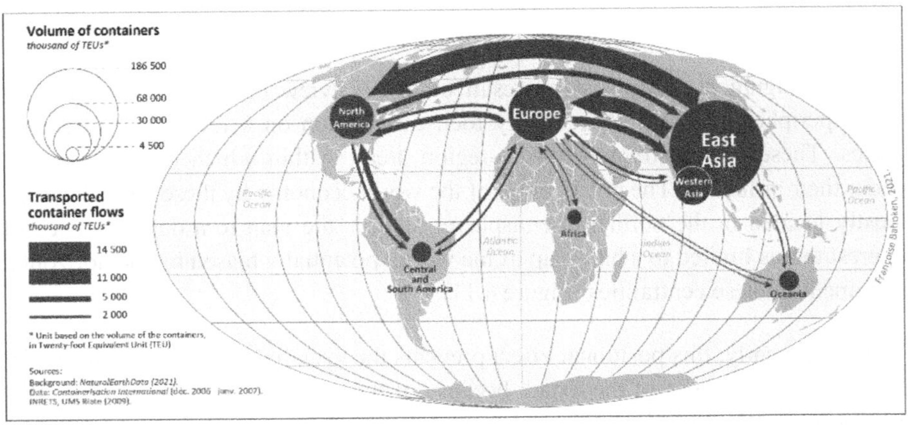

Figure 4.12. *The alignment effect of curved lines of global maritime flows (Bahoken 2021). For a color version of this figure, see www.iste.co.uk/mericskay/cartographic.zip*

The fact that the curved line is longer than the straight line may suggest that the distance traveled by these flows is greater. For example, the containers transported between the Asian regions and North America appear to have traveled a longer distance, which is not the case. Since the transition to the curve is, in practice, made by adding intermediate points, the result obtained could also suggest the presence of stops along the journey, such as port calls to carry out trans-shipment operations (loading/unloading), which do not exist in the data set.

The curve eventually leads to the transmission of erroneous information, introduced by the possible modification of the general organization of the transport, the increase in the trip length made in reality, which results in an increase in the cost of the trip, the transport time and so on.

This does not mean that one should avoid elaborating this style of flow maps. Rather, attention is drawn to the importance of the shape of the flow line in the understanding of the map, which a purely graphical approach distorts, especially at the global level, which is sensitive to the relative positions of places in relation to one another.

One could then think that this alignment problem could be largely solved by choosing a projection system that is appropriate to the theme to be represented, avoiding the need to curve the lines. Figure 4.13, for example, applies a Bertin projection[17] that repositions the different centralities and modifies the lines accordingly.

The variation in the positions of the locations, following a change in projection, has an effect on their spacing and, consequently, on the length of the flow lines. The change in projection alone is not enough to generate a "more accurate" image, the primary problem being the shape/morphology of the lines. To remedy this, the use of straight lines (more generally, geodesics) allows an expression of the flows as close as possible to the mobilized data, in other words, in the Euclidean metric, for a representation according to a flow logic (Figure 4.14).

17 This projection, conceived by Bertin in 1953, is presented by the author as a "regionally compensated projection in which the compensation is no longer homogeneous, but is sought in a greater deformation of the oceans, for the benefit of a lesser deformation of the continents" (Bertin 1967).

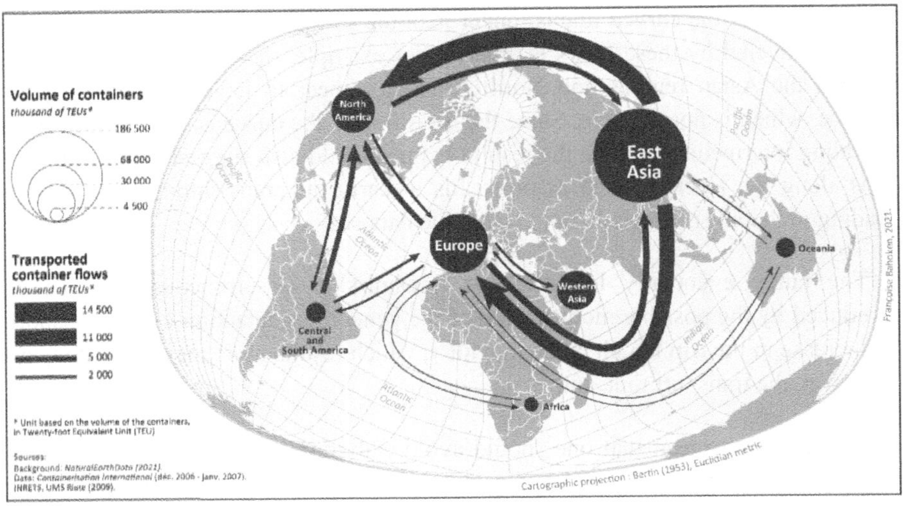

Figure 4.13. *Globalization of trade flows (curvilinear variant) (Bahoken 2021). For a color version of this figure, see www.iste.co.uk/mericskay/cartographic.zip*

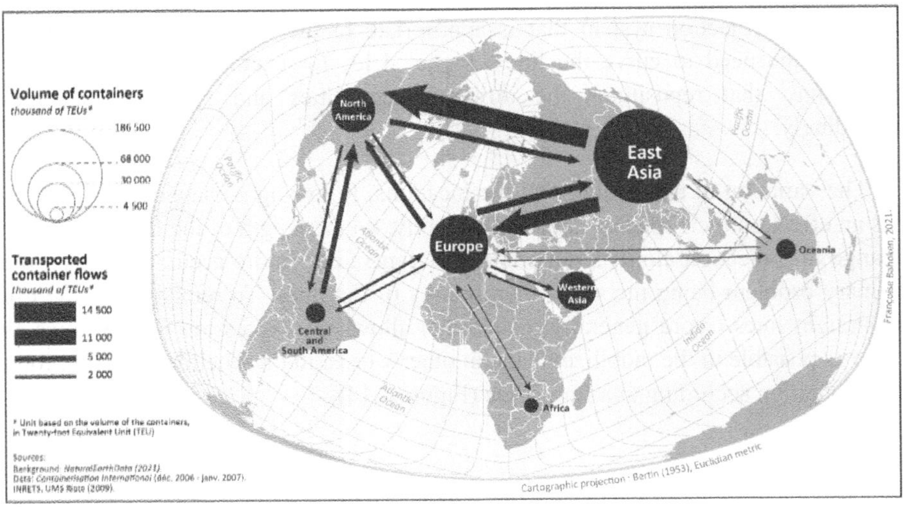

Figure 4.14. *Globalization of trade flows (rectilinear variant) (Bahoken 2021). For a color version of this figure, see www.iste.co.uk/mericskay/cartographic.zip*

The flows mapped above describe container transport between global macro-regions, using a mode of transport that can reasonably be assumed to be maritime.

Moreover, these flows are in fact subject to a cartography constraint linked to the alternation between land and sea, which means that the "route effect" must be taken into account.

4.4.1.2. *The route effect and the sinuosity of the lines*

The "route effect" requires that we follow a route in a given space (a maritime route, for example) and consider the geography of the spaces we cross in relation to the movement in question. Not taking this space of interest into account means taking the risk of a map describing a situation that is impossible in reality (for example, ships crossing continental spaces). Two cases are therefore possible, depending on whether the route taken is accurately known or not, and this information on the route is generally disconnected from information on the quantities transported (Figure 4.15).

Although the maritime routes used by ships are well known and can be geo visualized in real time on the web, they only describe the density of maritime traffic, not their semantics. The representation (plot) of maritime routes from AIS data proposed by AISHub for land antennas, Vessel Finder or Marine Traffic provides information on the density of the lines and potentially on their frequency according to the type of vessels. However, these data do not provide information on the content or quantities carried by the vessels, except when they can be matched with other databases that allow them to be characterized, such as the IHS Markit Lloyd's Fairplay vessel database[18], which is the most complete (Kerbiriou and Serry 2021). These lines therefore only shed light on traffic density. For this reason, these maps are built using link logic, where the lines have a unique width (Figure 4.15).

In the second case, where the trade route is unknown, two additional choices are possible. The first is to ignore the geography of the spaces crossed by these ships by representing only the transfer of these quantities of containers, as before. The second is to consider the geography of this transport but only in an approximate way, since the precise information is unknown.

It should be noted that it is easier to approximate the charted trajectory of vessels whose quantities, frequencies, flags, etc., are known than to estimate the quantities from a collection of precise spatiotemporal trajectories.

---

18 The Lloyd's Vessels database provides information to characterize all vessels with a temporal depth, namely: IMO and MMSI identification numbers of the vessel, type, size, capacity, owner and operator (Kerbiriou and Serry 2021).

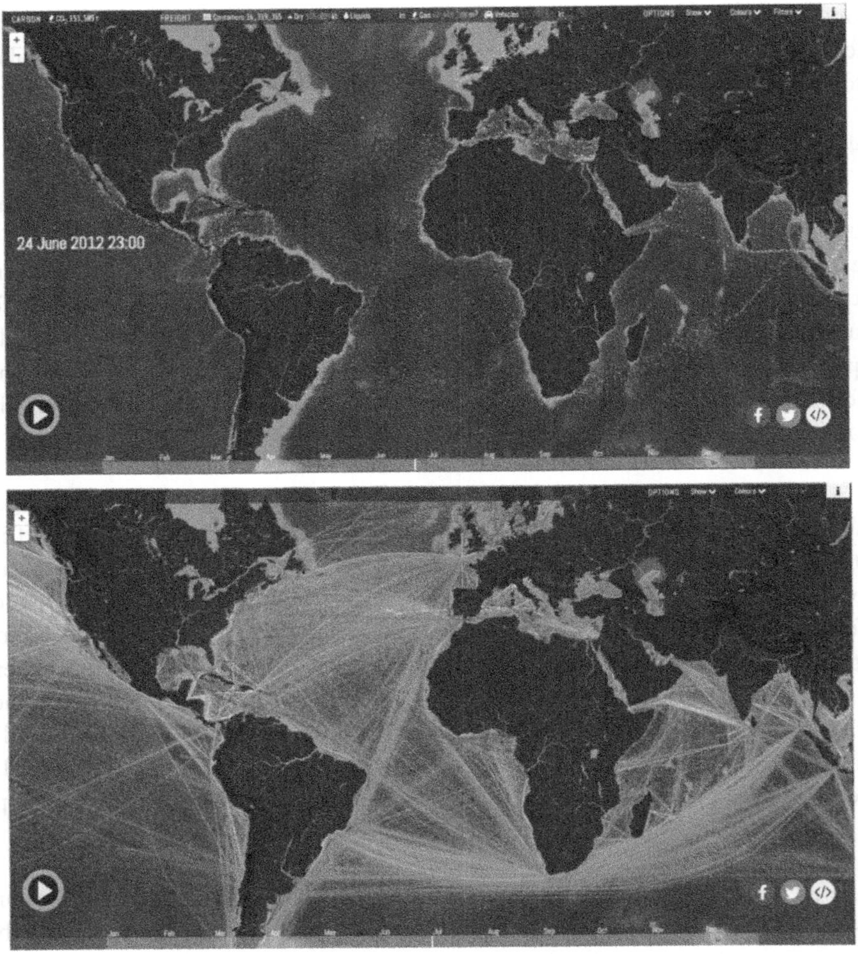

Figure 4.15. *Density of global maritime traffic (source: Shipmap.org). For a color version of this figure, see www.iste.co.uk/mericskay/cartographic.zip*

4.4.2. Geographical space in the representation of global flows

Not considering the geography of OD locations in the cartographic representation of global flows (or at other geographic scales) does not pose a problem in itself since the cartographic method that is implemented according to flow logic does not consider the space of positions in the representation. The fact is that space is only mobilized in the background map.

In a flow map, the territorial division generally occurs when the map is projected (via the spatial layout) to contextualize the signs and help with interpretation. It is therefore possible to remove this background map without changing the nature of the information perceived (Figure 4.16).

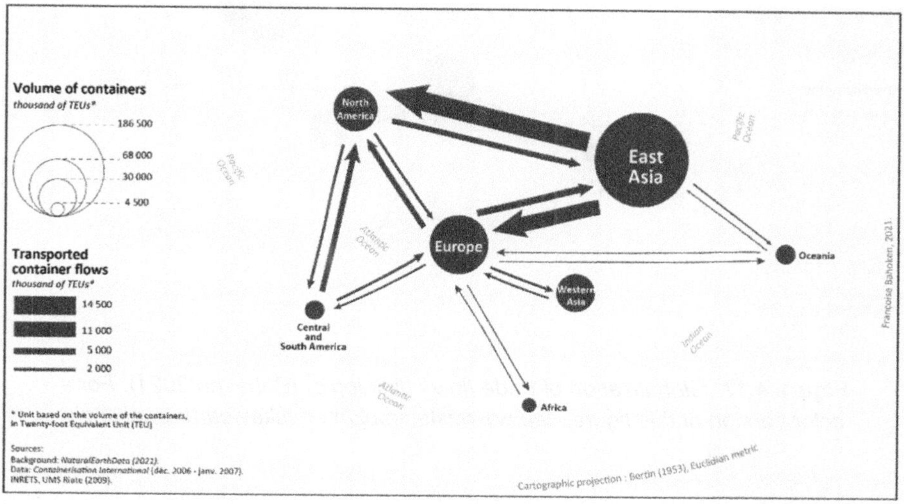

Figure 4.16. *Globalization of trade flows (flow logic) (Bahoken 2021). For a color version of this figure, see www.iste.co.uk/mericskay/cartographic.zip*

As the background map has been removed, it is no longer necessary to keep the geographical positions of the locations. The suppression of the geography allows a change of space at the limit, namely, the projection of these macro-regions in a geometrical space that is not necessarily geographical.

On the theoretical level, we move from a flow logic (gravitational logic) to a link logic (with reference to graph theory, social network analysis), which considers an adjacency matrix[19] where what is important, above all, are the positions of the vertices (translating the locations) in the plane (Figure 4.17).

19 An adjacency matrix is a binary flow matrix: cells are filled with 1's and 0's, signifying the existence or absence of a link between ODs, respectively.

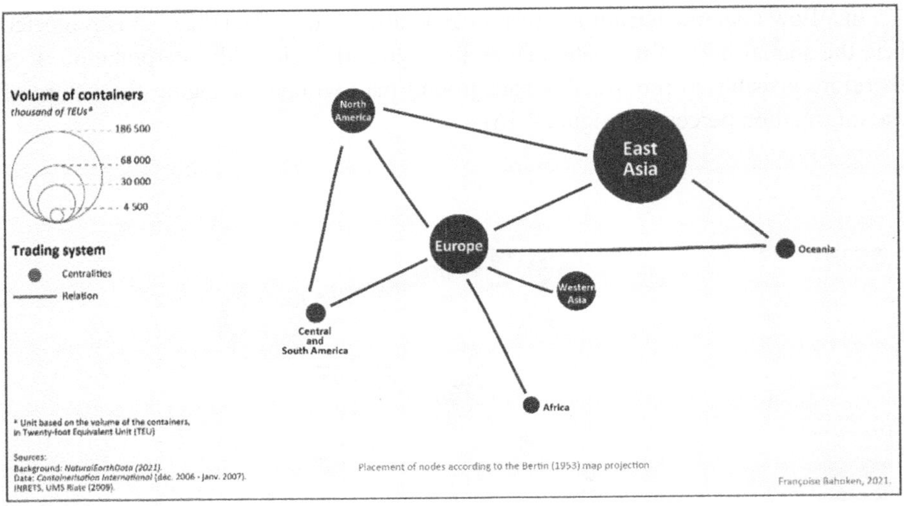

Figure 4.17. *Globalization of trade flows (link logic) (Bahoken 2021). For a color version of this figure, see www.iste.co.uk/mericskay/cartographic.zip*

In other words, the representation is no longer a map but rather a graph modeling the global container trade system. The links in this spatialized graph can also be weighted to enrich the representation with knowledge of the quantities of containers, which is supplemented and enhanced by indication of the direction (or orientation) of the vessels (Figure 4.16), or even by the color according to the type of vessel, its flag and so on.

Within the framework of graph theory, applying a logic of links leads to the repositioning of the locations of this maritime trade system to communicate on its structuring at the global level, on the morphology of the relational system (Figure 4.18), and no longer on their geography, which becomes secondary.

This redistribution of elements modifies the nature of the representation of these flows, the type of document (flow map/flow graph) and the nature of the information communicated: relational/structural/geographical. From a map to a graph, this last image relates more to computer graphics, offering the reader a simplified and schematized vision of world trade flows. The general morphology of this containerized trade system is present on a double geographical scale: between the major poles and a secondary regional network (Brazil, Russia, India, China and South Africa [BRICS]).

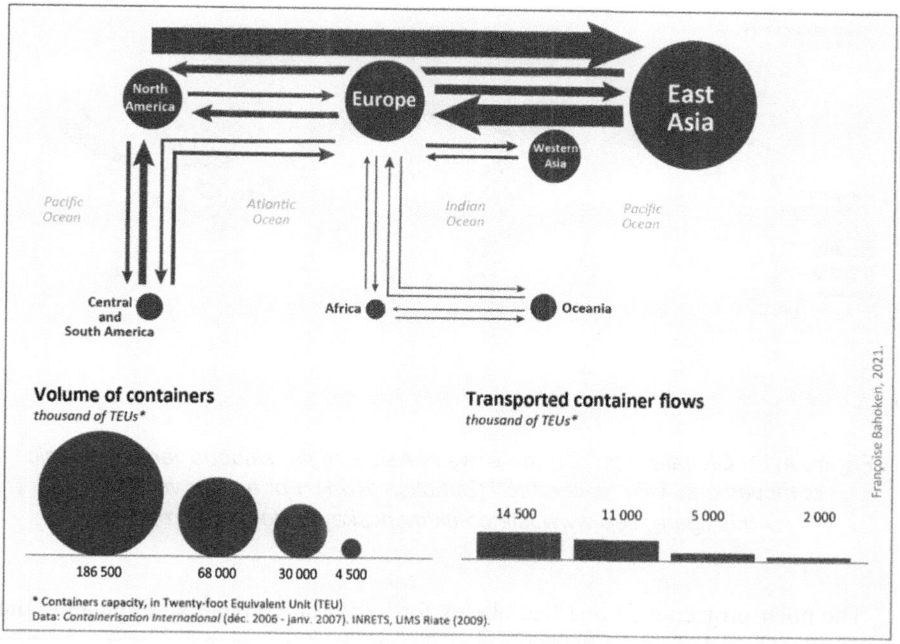

Figure 4.18. *Structure of containerized maritime transport (Bahoken, 2021). For a color version of this figure, see www.iste.co.uk/mericskay/cartographic.zip*

4.4.3. *Approximating space in the representation of global flows*

Rather than omitting geographical space completely from the representation of these flows, it is possible to reintroduce it in an approximate or imprecise manner, given that we do not have exact knowledge of it. To do this, it is necessary to adopt a line of thought according to a "logic of movements". A final way of representing these containerized movements from a geographical cartographic perspective is to constrain the expression of the plots in maritime space to bypass the land (to respect the "route effect").

To do this, one must "simply" integrate the "route effect" into the cartographic process, which would lead to a change in reasoning and to the adoption of a logic of movements. Any other way, the map will suffer from the "joint effect" (Figure 4.19).

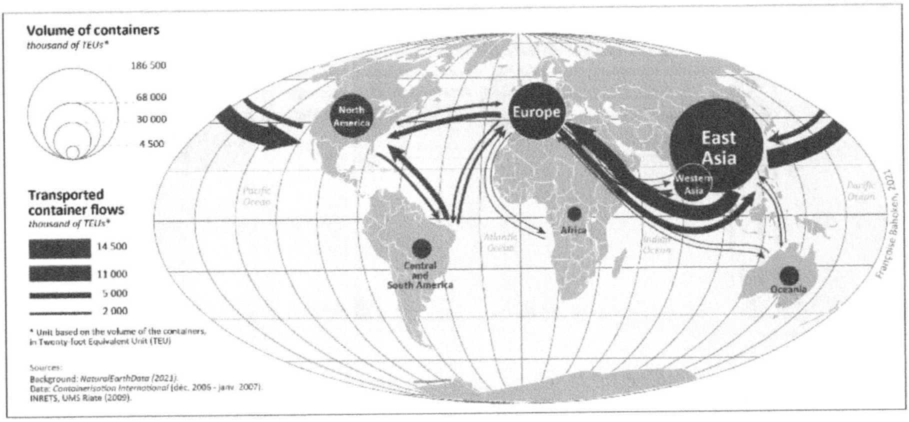

Figure 4.19. *Globalization of trade flows of Asian origin (sinuous variant -- Logic of movements with "joint effect" (Bahoken 2021). For a color version of this figure, see www.iste.co.uk/mericskay/cartographic.zip*

The polar projection is one that allows for territorial continuity, eliminating the "joint effect", including for maritime movements. It allows a spatial representation of all these flows while minimizing distortions (Figure 4.20).

The objective of visualizing the influence of central or major players on a world map has been dominant for many years and continues to be so in certain cases. However, the objective of communicating one's mastery of the world through a map is so prevalent that cartographers have been able to characterize certain methodological problems by directly associating them with the necessity to represent world powers.

For example, by referring to the cartographic plate he published in a *Fortune* atlas as the "American/British problem", the cartographer Harrison (1940) shows that he perceives the "route effect" as an impossibility with regard to exposing the scope of spatial coverage of the economic powers of the moment.

The consideration of space in the cartographic representation of maritime movements logically limits their representation through the necessity of their spatialization in maritime space in order to take into account global-level cartographic constraints. Only a detailed knowledge of the routes associated with the quantities transported would make it possible to acquire a precise representation of the reality of this transport, which, as mentioned before, is generally not possible.

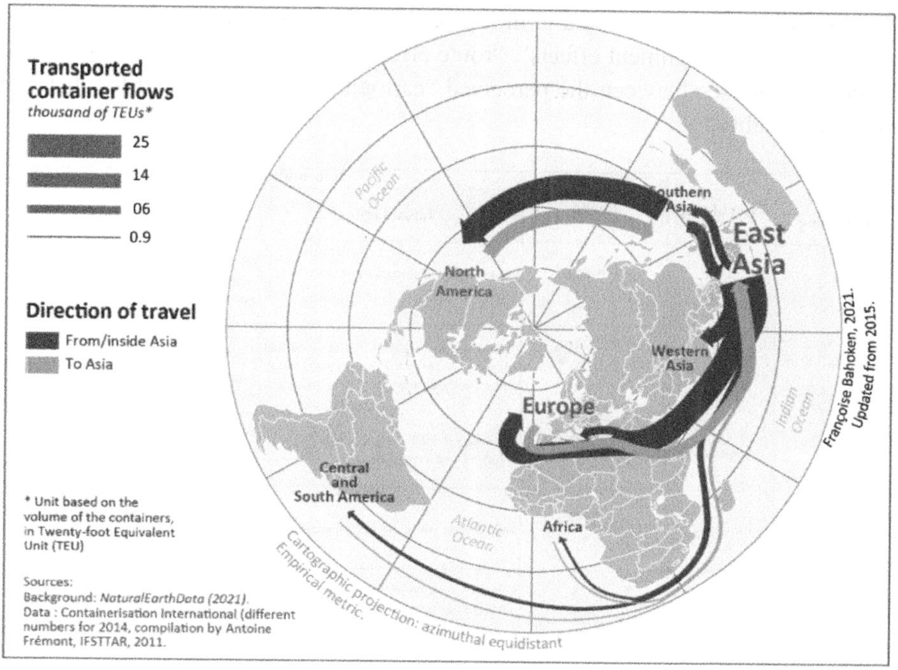

Figure 4.20. *Globalization of trade flows of Asian origin (sinuous variant, logic of movements in polar projection) (Bahoken 2021). For a color version of this figure, see www.iste.co.uk/mericskay/cartographic.zip*

4.5. Conclusion

Although produced with the same data, the maps presented in this multi-representation exercise differ in their apparent design, which more or less accurately incorporates the main challenges of mapping these trade flows. Figure 4.21 presents a first summary table, comparing the consideration of the specificities of flow mapping and graphing these global flows in the different versions proposed.

Each of the maps (in rows) is associated with three families of features (in columns):

– the level of graphic perception linked to the perception of complexity, which is itself related to the represented information or its graphic translation;

– the general challenges of flow mapping such as the “spaghetti effects” and “amalgam effects”;

– specific challenges related to the presentation of a geographical context, to the consideration of “alignment effects”, “route effects” and “joint effects”, as well as to the correspondence between the perceived “cartographic distance” and the distance traveled in reality.

Type of reasoning	Type of map	Image	Information provided	
			Advantages	Disadvantages
Geographical	flowmap	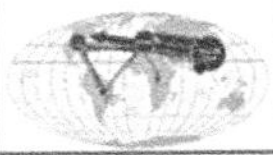	View of a general geography of the trading system.	Formally imperfect visualization. Confusion in the perception of the quantities exchanged. Amalgam effect.
	flowmap		Clear perception of the quantities exchanged and expressed at two different geographical scales. Aesthetic improvement by the curves.	Imperfect visualizations though very aesthetic patterns. Flow information distorted by curvature of lines which artificially increases perceived map distance and modifies overall transport organization.
	flowmap	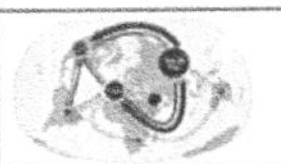	Distinction of the Triad and a center-periphery spatial organization.	
	flowmap		Most correct view in the Euclidean metric.	Spatial issues at world level not taken into account.
	movement map		General perception of the geography of maritime transport movement. Reconciles flow mapping with maritime transport.	“Approximate” representation of maritime routes.
	movement map		Accurate perception of the geography of a sector (oil transport). Accurate representation of routes and quantities.	Approach by economic sector difficult to generalize to a set of transported goods.
Morphological	graph	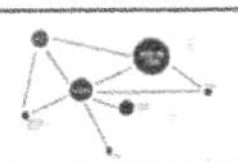	Accurate perception of the structure of the maritime transport system: of the hierarchy of locations and quantities exchanged.	Absence of transport geography (except for reading labels). Lack of perception of quantities.
	graph of flows/flow graph		Most interesting perception of the morphology of the transport system analyzed. The geography is perceptible through the locations labels.	Computer graphics that are no longer cartographic. Elude the geography of trade in favor of the morphology of the system.

Figure 4.21. *Summary of the management of global flow mapping challenges (Bahoken, 2021). For a color version of this figure, see www.iste.co.uk/mericskay/cartographic.zip*

We thus observe that the level of overall graphic perception is very good in two opposite cases: when the image integrates the geographic space in the cartographic drawing of the flow line and when the latter is not taken into account.

The global maritime movements mapped as a sinuous line approximating the maritime routes, and according to a cartographic projection that respects their spatial continuity, present a very good level of graphic perception. While the failure to consider the “joint effect” puts forward a problem of thematic interpretation, suggesting an interruption in the transport operation, the cartographic representation is considered to have a good level of graphical perception because it suggests the routes used.

The level of graphical perception is also very good when these same movements disregard geographical space and challenges that are related to the perception of "cartographic distance". Their representation in the form of one spatial flow graph shows the morphology of the maritime trade system more than their geography, which is accessible at a second level of reading by the toponyms. Only one effect is not considered by these graphs, that of alignment, which is potentially solved by an adequate placement of the different locations. The efficiency in the representation of relational data in the form of a graph, linking localized or non-localized actors, depends on the judicious placement of the corresponding nodes in order to reveal centralities, structuring of communities (or not) and possible hierarchies between places.

The representations of these flows that seem the least effective, from our point of view, are those that do not fit well with geographical space, either because they evade it or because they overestimate it. In this case, an exaggerated vision of the "cartographic distance" is proposed, linked to stylistic tricks and leading to the curve of the line/arrow. Although the images obtained are, indeed, more esthetic and more pleasant to look at than those using straight lines, the problem is that they are also not very meaningful. They communicate information that is all the more erroneous the larger the graphic distortion by providing an excessive distortion of the represented information (it should also be noted that their level of graphic perception of these images, which overestimates the space, is considered average).

Thus, in the absence of being able to correctly estimate the routes taken in reality, rather than using esthetic devices that give the illusion of representing movement, it is preferable to underestimate the role of geographic space by using the Euclidean metric, implemented in a projection that is adequate for the scale represented. The level of graphical perception will nevertheless be good because only the "route effect" will not be considered.

The apparent diversity of the proposed series of maps is due to the way in which the different issues are considered. The varied expression of the resulting cartographic designs depends, in fact, on the underlying theoretical reasoning from which the mentioned (cartographic) considerations result. Figure 4.22 presents a second summary table, where the different map images participate in geographic (gravity) or morphological (reticular) reasoning. Their conceptual differences are reinforced by the way they consider the space of this containerized maritime trade system. These two main families of reasoning are characterized in columns by a type of image, a flow/movement map and a flow graph with advantages and disadvantages.

Image	Level of graphic perception	General issues		Spatial issues at world level				
		Spaghetti effect*	Amalgam effect	Geographical context	Alignment effect	Route effect	Joint effect	Cartographic distance
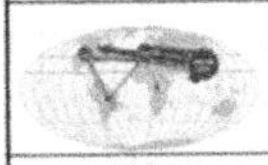	Low	Yes	Yes	Yes	No	No	Yes	Underestimated
	Medium	No	No	Yes	No	No	Yes	Overestimated
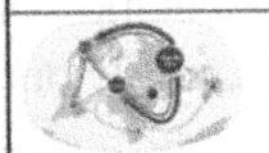	Medium	No	No	Yes	Yes	No	Yes	Overestimated
	Good	No	No	Yes	Yes	No	Yes	Underestimated
	Good	No	No	Yes	Yes	Yes	No	Approximate
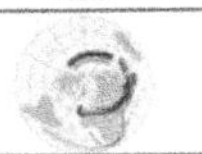	Very good	No	No	Yes	Yes	Yes	Yes	Approximate
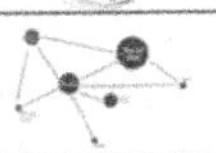	Good	No	No	No	No	No	NC	Eluded
	Very good	No	No	No	Yes	No	NC	Eluded

* The spaghetti effect is by definition small for this level of aggregate flows in macro-economic regions.
NC : Not concerned.

Figure 4.22. *Advantages and disadvantages of the different flow/movement maps (Bahoken, 2021). For a color version of this figure, see www.iste.co.uk/mericskay/cartographic.zip*

The map and the graph of flows have their advantages and disadvantages in terms of graphically communicating information about these flows. Beyond their differences in terms of graphically translating a message, these two forms of (carto)graphic representation are not opposed but complementary in terms of communication. They lead to different visions of these exchanges and allow several interpretations, depending on the choice of the type of representation. It is, above all, the cartographer's intention to either show the general structure of the system of exchanges (thanks to computer graphics) or to show the geography of this same system, represented with a traditional or polar projection, for example, to draw attention to commercial centralities (such as the Triad).

Although graphs and maps are part of complementary reasoning, the challenges of their representation also raise questions of rhetoric related to the reception of the map's message by a given audience. Indeed, beyond the theoretical and methodological aspects, as well as the overall design, the mapping of global flows, like other types of maps, is a function of the theme of transfers and exchanges between ODs.

The perception of exchanges varies according to whether the medium is static or interactive. The existence of new possibilities of dynamic, interactive or animated geovisualization of flows/movements leads to a renewal of the modalities of representation, especially the modes of interaction between users, maps and data. However, the actual possibilities still do not integrate their reception, which again raises the question of the rhetoric of the map in relation to the tools used.

Although the "spaghetti effect" is the first problem that must be solved to obtain a comprehensible and interesting representation, it is also the most important one since it concerns all maps of flows and movements, regardless of their scale, as soon as they include more than 10 pairs of OD locations. Other effects can then appear, depending on the theme in question and the cartographer's intention. Economic movements describing maritime circulations do not call upon the same cognitive, social and geopolitical drivers (Bacon et al. 2016) as those describing demographic movements. Even if the graphic signs used (points, lines/arrows) are similar, the fact that their semantics and interpretations differ requires particular attention to their representations.

The multi-representation work proposed here has focused attention on the questions surrounding the mapping global trade flows, thus highlighting the role of specific spatial challenges (related to "alignment", "route" or itinerary and "joint" effects). If we had dealt with other types of flows and movements, we would have been confronted with other issues for their representation. The case of migratory movements, for example, induces specific problems that are, above all, linked to the geopolitical context of the moment and to humanistic considerations related to human rights or political ideologies, imposing particular attention on the rhetoric and the style of the map: the choice of words, forms, colors, and signs.

At a time when data on flows and movements are multiplying, particularly around transport, mobility and migration, it is important to question and investigate this form of cartography, which is playing an increasingly important role in the practices of both traditional players (cartographers, geographers) and newcomers (web designers, private platforms, data journalists, NGOs, etc.) The cartography of flows and movements requires now, more than ever, to be understood according to informational, technical, methodological and stylistic perspectives in order to reflect and propose effective forms of cartographic communication adapted to the representation of a world that is moving.

4.6. References

Amilhat Szary, A.L. (2018). *Qu'est-ce qu'une frontière aujourd'hui ?* Presses Universitaires de France, Paris.

Bacon, L., Clochard, O., Honoré, T., Lambert, N., Mekdjian, S., Rekacewicz, P. (2016). Cartographier les mouvements migratoires. *Revue européenne des migrations internationales*, 32(3/4), 185–214.

Bahoken, F. (2014a). Pour le renouvellement de la sémiologie de la carte de flux. *Cartes et Géomatique*, 222, 109–121.

Bahoken, F. (2014b). Éléments pour une représentation (carto)graphique des matrices de flux. *Mappemonde*, 115 [Online]. Available at: http://mappemonde.mgm.fr/num43/articles/art14304.html [Accessed 12 July 2021].

Bahoken, F. (2016a). L'approche cartographique de la décomposition des matrices de flux. *Mappemonde*, 116 [Online]. Available at: http://mappemonde-archive.mgm.fr/num44/articles/art14404.html [Accessed 12 July 2021].

Bahoken, F. (2016b). Contribution à la cartographie d'une matrice de flux. PhD Thesis, Université Denis Diderot, Paris.

Bahoken, F. (2019). L'évolution de l'image cartographique du flux au prisme de l'analyse des interactions spatiales. *Siècles* [Online]. Available at: http://journals.openedition.org/siecles/4142 [Accessed 15 Feb 2023].

Bahoken, F. and Lambert, N. (2020). Méfiez-vous des cartes, pas des migrants. *AntiAtlas journal* [Online]. Available at: https://www.antiatlas-journal.net/04-mefiez-vous-des-cartes-pas-des-migrants/ [Accessed 15 Feb 2023].

Bahoken, F., Grasland, C., Zanin, C. (2016). D'une cartographie de flux à une cartographie du mouvement. Aspects sémiologiques. *Cartes et Géomatique*, 229/230, 65–74.

Bahoken, F., Jégou, L., Lagarde, D., Lambert, N. (2020). La séduction des cartes du geoweb. Le cas des flux de migrants internationaux. *Cybergeo: European Journal of Geography* [Online]. Available at: http://journals.openedition.org/cybergeo/33792 [Accessed 15 Feb 2023].

Bertin, J. (1967). *Sémiologie graphique. Les diagrammes, les réseaux, les cartes*. Gauthier-Villars-Mouton, Paris.

Boyandin, I. (2010). Using flow maps to explore migrations over time. In *Proceedings of the International Conference on Geographical systems.* AGILE 2010, Guimarães.

Boyandin, I., Bertini, E., Bak, P., Lalanne, D. (2011). Flowstrates: An approach for visual exploration of temporal origin-destination data. *Proceedings of the Computer Graphics Forum*, 30(3), 971–980.

Cauvin, C. (1997). Au sujet des transformations cartographiques de position. *Cybergeo: European Journal of Geography.* [Online]. Available at: http://journals.openedition.org/cybergeo/5385 [Accessed 13 July 2021].

Cauvin, C. (2020). Waldo Tobler. *Hypergéo* [Online]. Available at: https://www.hypergeo.eu/spip.php?article763 [Accessed 12 July 2021].

Cauvin, C. and Reymond, H. (1991). Interaction spatiale et cartographie : les solutions de W. Tobler. *Espace, populations, sociétés*, 3, 467–485.

Claudel, M., Nagel, T., Ratti, C. (2016). From origins to destinations: The past, present and future of visualizing flow maps. *Built Environment*, 42(3), 338–355.

Côme, E., Bapaume, T., Jégou, L., Bahoken, F., Maisonobe, M., Roelandt, N., Le Campion, G. (2019). Arabesque, application d'exploration et de géovisualization de données de flux et de réseaux. In *Actes de la conférence internationale de géomatique (SAGEO'2019).* Sage, Clermont-Ferrand.

Côme, E., Bahoken, F., Lambert, N. (2021). Winds, a tribute to Tobler. In *Recueil des contributions au Colloque Cartomob*, Toulouse.

Didelon, C. (2013). Le Monde comme territoire ; pour une approche renouvelée du monde en géographie. HDR, Université de Rouen, Rouen.

Didelon, C., Grasland, C., Richard, Y. (2008). *Atlas de l'Europe dans le monde.* La Documentation française, Paris.

Grasland, C. (2010). Spatial analysis of social facts, a tentative theoretical framework derived from Tobler's first law of geography and Blau's multilevel structural theory of society. In *Handbook on Theoretical and Quantitative Geography*, Bavaud, F., Mager, C. (eds) [Online]. Available at: https://halshs.archives-ouvertes.fr/halshs-00410669v2 [Accessed 21 July 2021].

Grootens, J. and Rendgen, S. (2020). *Le système Minard : anthologie des représentations statistiques de Charles-Joseph Minard.* B42, Paris.

Holten, D. (2006). Hierarchical edge bundles: Visualization of adjacency relations in hierarchical data. *Proceedings of the IEEE Transactions on Visualization and Computer Graphics*, 12(5), 741–748.

Jenny, B., Stephen, D.M., Muehlenhaus, I., Marston, B.E., Sharma, R., Zhang, E., Jenny, H. (2018). Design principles for origin-destination flow maps. *Cartography and Geographic Information Science*, 45(1), 62–75.

Joliveau, T. (2010). Le géoweb, un nouveau défi pour les bases de données géographiques. *L'Espace géographique*, 2(40), 154–163.

Kerbiriou, R. and Serry, A (2021). Les signaux AIS et la cartographie de la circulation maritime. In *Recueil des contributions au Colloque Cartomob*, Toulouse.

Lambert, N., Bahoken, F., Côme, E. (2021). Vers une relecture du Flowmapper. In *Recueil des contributions au Colloque Cartomob*, Toulouse.

MacEachren, A.M. (1995). *How Maps Work: Representation, Visualization and Design.* The Guilford Press, New York.

Mélançon, G. (2010). Living flows: Enhanced exploration of edge-bundled graphs based on GPU-intensive edge rendering. In *Proceeding of the 14th International Conference Information Visualisation*. IEEE, London, 523–530.

Mericskay, B. (2016). La cartographie à l'heure du géoweb : retour sur les nouveaux modes de représentation spatiale des données numériques. *Cartes et Géomatique*, 229/230, 37–50.

Nystuen, J.D. and Dacey, M.F. (1961). A graph theory interpretation of nodal regions. *Papers and Proceedings of the Regional Science Association*, 7(1), 29–42.

Palsky, G. (1996). Des chiffres et des cartes. Naissance et développement de la cartographie quantitative française au XIXe siècle. Comité des travaux historiques, Paris.

Phan, D., Xiao, L., Yeh, R., Hanrahan, P., Winograd, T. (2005). Flow map layout. In *Proceedings of IEEE Symposium on Information Visualization*. INFOVIS 2005', Minneapolis, 219–224.

Plantin, J.C. (2012). Les pratiques de cartographie numérique en ligne. Expression, remédiation, circulation. PhD Thesis, Université de Technologie de Compiègne, Compiègne.

Poulain, M. (1981). Contribution à l'analyse spatiale d'une matrice de migrations internes. PhD Thesis, Université de Louvain La Neuve, Cavey.

Ravenstein, E.G. (1885). The laws of migration. *Journal of the Statistical Society of London*, 48(2), 167–235.

Ravenstein, E.G. (1889). The laws of migration, second paper. *Journal of the Statistical Society of London*, 52(2), 241–305.

Thornthwaite, C.W. (1934). Internal migration in the United States, Bulletin number one. Study of population redistribution. Report, Wharton School of Finance and Commerce, Industrial Research Department, Philadelphia.

Tobler, W.R. (1970a). A computer movie simulating urban growth in the Detroit region. *Economic Geography*, 46(2), 234–240.

Tobler, W.R. (1970b). Experiments in migration mapping by computer. *The American Cartographer*, 14(2), 155–163.

Tobler, W.R. (1979a). Estimations of attractivities from interactions. *Environment and Planning A*, 11(2), 121–127.

Tobler, W.R. (1979b). *A Geographical Flow Mapping Program.* University of California, Santa Barbara.

Tobler, W.R. (1981). A model of geographical movement. *The Geographical Analysis*, 13(1), 1–20.

Tobler, W.R. (1982). Cartographic study of movement tables. In *Presentation in the Mapping National Computer Graphics, Session on Statistical Graphics*, Anaheim.

Tobler, W.R. (1987). Experiments in migration mapping by computer. *The American Cartographer*, 14(2), 155–163.

Tobler, W.R. (1995). Migration: Ravenstein, Thorntwaite, and beyond. *Urban Geography*, 16(4), 327–343.

Tobler, W.R. (2012). *Migrations: Maps and Models.* University of California, San Bernardino.

5

Cartographic Communication on the Geoweb: Between Maps and Data

Matthieu NOUCHER
Passages, CNRS, University of Bordeaux, France

5.1. Introduction

In his reference work *The Sovereign Map*, Jacob, like many other authors before him (Robinson 1952; Moles 1964) inscribes the map in the communication process: "A map is defined perhaps less by formal features than by the particular conditions of its production and reception, by its status as an artefact and as mediation in a social communication process" (Jacob 1992, p. 41). By using the map as an intermediary between the transmission and reception of information, he anchors his remarks in the perspective proposed by Shannon (1948) with his Map Communication Model (MCM). The explicit aim of this model was to quantify the probability of message transmission and to reduce the "noise" that interferes with the transmission of information. The Czech geographer Koláčný would establish this model as a reference for thinking about map communication throughout the 1970s (Plantin 2014).

Koláčný's model (Figure 5.1) complicates Shannon's original diagram by de-multiplying the sub-processes of communication, both at the level of information encoding by the mapper and the decoding by the receiver. In particular, he emphasizes the importance of several parameters: the "goal" (11), "empirical knowledge" (12), "capacities" (13), "psychological processes" (14) and "external conditions" (15). He also incorporates the map reader's "feedback" (7).

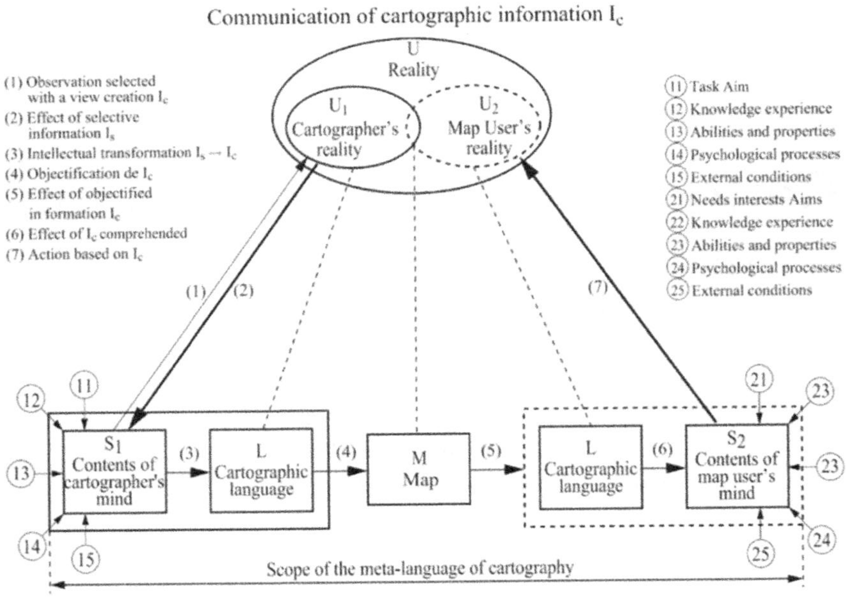

Figure 5.1. *Koláčný's model of cartographic communication (source: Montello (2002) according to Koláčný (1969))*

As Plantin (2014) points out, Koláčný's appointment as head of the working group on cartographic information within the powerful American learned society, known as the International Cartographic Association, undoubtedly helped to secure this model as a dominating paradigm of cartographic communication until the 1980s. Koláčný's model, based on the MCM, is based on four assumptions: there is a clear separation between the cartographer and the user of the map; the map serves as an intermediary between the two; the map communicates information from the cartographer to the user; and cognitive and psychophysical parameters determine the user's ability to understand, learn and remember the information communicated by the map (Crampton 2001).

From the 1980s onwards, this dominant model was the object of strong criticism. Bougnoux (1991), for example, denounced the blind spots of this analytical framework, which overlooked the social and collective nature of communication: not only is the map a tool for transmitting information but it also appeals to emotions and collective consciousness. The development of geographic information systems (GIS) during the 1980s led to a shift in communicational analyses, which were gradually abandoned (Montello 2002) in favor of new approaches that moved away from the behavioral drifts of the MCM.

Among these research trends, an alternative program has emerged, moving away from discourses on map efficiency and instead focusing on their social and political uses. It is called critical cartography and proposes a critical reading of cartographic production in history, highlighting the socially constructed aspect of the map, and its capacity to serve strategies and relations of knowledge/power. Distancing themselves from the modernist and evolutionary narratives that have characterized the history of cartography until now, researchers such as Harley and Wood question the political as well as the social and contextual aspects of cartographic commissioning and production. Opposing positivist approaches that see maps as mirrors of nature (Harley 1989), critical cartographers instead attempt to reveal the intentions that guide their mapping[1], as well as the performance generated by their uses (Wood 1992). This chapter is an extension of this critical approach[2] to cartography, focusing on the emerging challenges of new practices, which are linked to the development of the Geoweb.

With the digital revolution, these analyses have become even more complex. The acceleration of the "digitization process of our societies" (Cardon 2015) is now marked by the proliferation of devices that are dedicated to the production or consultation of spatial data. The development of geomatics, the increasing availability of spatial data, the growing involvement of digital multinationals and, in a second step, the rapid growth of Internet mapping combined with the emergence of collaborative practices have contributed to the advent of the Geoweb (Kitchin et al. 2017). This can be defined, *a minima*, as the aggregation of spatial technologies and geo-referenced information that is organized and transmitted by the Internet. Maps are now everywhere on the web. How, then, can we rethink the cartographic communication process in light of these reconfigurations? If the abundance of online maps testifies to the propensity of today's societies to privilege them as the main descriptors of the world, how can we rework the communication processes through maps?

---

1 French lacks the equivalent of the English word *mapping* (Kitchin and Dodge 2007) to describe conceptual, methodological and instrumental tools, as well as the measure of action, of doing, and of performance, that is necessary to map the world and give it meaning. We then proposed the expression "*fabrique cartographique*" (Joliveau et al. 2013).

2 Readers interested in both the history and contemporary challenges of critical cartography are invited to consult another volume from the ISTE/Wiley encyclopedia that is coordinated by Debarbieux and Hirt (2022), and which is dedicated to the interactions between cartography and politics.

In an attempt to answer these questions, this chapter examines the four assumptions that have been previously mentioned, with regard to the new ways of making and using maps in the Geoweb era. The first part emphasizes the technical, legal and organizational reconfigurations generated by the digital revolution. This overview highlights two key characteristics of the Geoweb: the malleability and the individualization of data, which are shaking up cartographic communication. Taking these two characteristics as a key, the second part questions the (now artificial?) separations between professionals and amateurs, producers and users of maps. The conclusion ultimately considers some perspectives for rethinking the links between communication studies and geographic information sciences.

5.2. Is Geoweb a new word for a new reality?

"Information explosion and technological revolution": this is how Bédard (1993) described the impact of geomatics development in organizations in the early 1990s. Since then, geo-information technologies (GIT)[3] have been democratized under the effect of the growing expansion, first of micro-computing, then of the Internet. This technological revolution has been accompanied by an information explosion, initially from geomatics experts to thematic specialists (planners, geologists, foresters, urban planners, hydrologists, etc.) and then, more recently, from the general public, which now handles geographical information on a daily basis (Joliveau 2020). The Geoweb therefore appears less as a technological revolution than as a series of technological, legal, and organizational evolutions, which are certainly dazzling, but are also logical, and which raise numerous questions, particularly in terms of cartographic communication.

5.2.1. *A slow and progressive opening of the expert approach of geomatics*

Cartography and GIS have long remained a matter for professionals or enthusiasts, although a combination of the two is not necessarily excluded. The first users of what was not immediately called GIS appeared at the end of the 1970s, at a time when computers were expensive, cumbersome and required computer science experts. Only large structures could afford to migrate their production of plans or maps from the drawing board to the digitizing table, and then to the screen (Feyt 2004). The end of the 1970s thus marked the development of computer-aided

3 GIT refers to geographic information systems, image processing tools, drawing instruments or computer-assisted cartography. We refer here to a definition that is not very restrictive, as proposed by Roche (2000).

drawing (CAD). Carried by this evolution and the will to automate update operations, the first urban databases were born from local initiatives, sometimes small scale, which were often isolated (Roche 2000). The first managers to embark on this path in France were essentially the technical services of major cities (Marseille or Lille, for example) and large network operators (EDF-GDF in particular), for whom the financial burden induced by the transition to digital technology weighed relatively little in relation to the overall costs and challenges of their activity. The way to what was originally perceived and conceived as computer-aided cartography (CAD)[4], then as GIS, was then opened.

After trials with the first microcomputers in the 1980s, the early 1990s marked the effective start of microcomputing, which had a decentralizing impact by allowing users to appropriate computer tools and to free themselves from the tutelage of computer scientists (Pornon 2004). In geomatics, this is the generation of desktop GIS such as ArcView, MapInfo, Geomedia and others, often acquired autonomously by services within the framework of projects with sectoral objectives. Thematic specialists from all backgrounds gradually took control of the tools, constantly broadening the fields of intervention, as shown by the multiplication of neologisms that appeared at that time: geomarketing, geodecisional, geobusiness, archaeomatics, geotraceability, geovisualization, etc.

In parallel to the diffusion of a culture of geographical information among managers in France, one must note the importance of geomatics in the field of defense, or even in the academic world. Henri Desbois explains how the militarization of cartography in the 19th century has allowed the defense sector to be at the forefront of geomatics issues, even today (Desbois 2015). Joliveau emphasizes the "interdisciplinary" nature of geography and computer science, which gives geomatics its scientific originality and its atypical positioning in the academic field, with different scientific networks: from the work of GIP Reclus[5] in the mid-1980s to the GdR MAGIS[6] today (Joliveau 2020).

This diffusion of geographic information technologies (from the military to scientists and managers) has led to an exponential increase in the amount of spatial data produced. Beyond the reference systems (topographic survey, photogrammetric plan, etc.) that fed the first urban databases, more and more thematic data are

4 The acronym CAD may have had several meanings. For example, Waldo Tobler distinguished in his courses between automatic cartography, which "automated" known cartographic procedures, and computer-assisted cartography, where cartography was somewhat designed and new techniques/methods were introduced.

5 Available at: www.mgm.fr/ARECLUS/gipf.html.

6 Available at: www.gdr-magis.imag.fr.

becoming the information base of many sectoral GIS (development, transport, environment, etc.). The ease of use of these tools, made possible by increasingly user-friendly interfaces, is accelerating their adoption by non-specialists. The deployment of GIS is then precipitated in organizations without any real concern for consistency or federation of tools and data. Individual cartographic production is often privileged. Centralized urban databases are giving way to the multiplication of locally produced and managed map layers. What users gain in autonomy, the organization loses in coordination. This dissemination of spatial data and maps, which is sometimes a bit messy, has nevertheless the merit of acculturating thematicians, and sometimes even decision-makers, to cartographic representations and geographical information.

5.2.2. *The Geoweb: a geospatial revolution?*

The emergence of web technologies will hasten the diffusion of geomatics beyond the professional sphere. Although online GIS appeared at the end of the 1990s, it is really only in the middle of the first decade of the 21st century that the opening to the non-expert accelerated. The initiatives of the web giants who, in the early 2000s, offered simple systems for producing and operating interactive digital maps were met with an enthusiastic response, both from the public and from thousands of producers of websites and mobile services. Following the acquisition of Keyhole Inc., Google released its first version of Google Maps in 2005 in the United States, then in Great Britain, before it appeared in France in 2006. This release coincided with the democratization of GPS chips, which were gradually being integrated into everyday objects (phones, watches, cars, etc.). In turn, these two movements produced a multiplication of geodata and practices around cartography. Anyone can now consult, annotate or cross-reference data on the web by georeferencing them, and using them to produce their own maps.

This had led to the emergence of many spontaneous uses in the cultural, artistic or political fields, but also others more solicited in the fields of geo-marketing, social networks, citizen science, crisis management or public participation (Joliveau 2010). The individual's positioning at the heart of spatial data production generates multiple uses, characterized by the emergence of a wave of expressions aimed at qualifying these evolutions: from user-generated geographic content (Purves et al. 2008) to geo-crowdsourcing (Sui et al. 2013) in order to emphasize the intrusion of Internet users in data production, and from GIS/2 (Miller 2006) to neogeography (Turner 2006), in order to highlight the disruptions with so-called "conventional" geomatics.

To translate these technical and organizational recompositions that are shaking up the world of geomatics under the influence of Web 2.0, Goodchild (2007) published a reference text in the field of geographic information science (GIScience). In it, he proposed the term "volunteered geographical information" or VGI.

This notion, as a unifying framework for heterogeneous practices, brings together all the approaches to creating geodata under a single term, both voluntary and spontaneous, implicit as well as explicit, which today provide spatial data that is different from, or even an alternative to, the conventional and institutional productions of professionals in the sector (Noucher 2019).

Reserved for experts yesterday, geomatics (or at least a certain form of it) is now becoming a mass phenomenon. Like the passage from Web 1.0 to Web 2.0[7], this new form of geomatics makes it possible for the largest number of people to consult a wealth of geographical information but also to publish, complete, annotate and combine digital maps, as well as to imagine new uses for them. While traditional geomatics, which originated in topography and surveying, seeks precision, rigor and legibility through a highly standardized approach carried out by professionals, Geoweb tools, which are used in everyday life (mobility, leisure, meetings, etc.), are more approximate and less elaborate, but accessible to a very large number of people (Joliveau et al. 2017).

The abundance and diversity of data accessible via these services could, however, compensate for their imprecision or disparity. Some authors argue, for example, that in the field of citizen science, the lower quality of opportunistic data (as opposed to protocolized data) could be compensated by its quantity (Crall et al. 2011). The simplicity and openness of the tools, as well as the multiplication of applications, could lead to the emergence of a posteriori knowledge and courses of action that an analytical approach would not be able to offer. The intervention of users at all levels would encourage the democratic appropriation of territorial issues in a renewed dialogue with experts and decision-makers. The Geoweb is also arousing significant interest in spatial data, among web designers and developers. The plethora of original, dynamic and interactive online maps that they produce broadens and renews traditional cartographic approaches and accustoms users to other aesthetic options as well as to other modes of cartographic representation[8].

---

7 Web 1.0 refers to the first generation of web interfaces, where the user was passive and could only read online articles. It became interactive and then contributive: with the appearance of forums, blogs and then social networks Web 2.0 thus emerged.

8 See Chapters 6 and 7.

5.2.3. *The map's place*

The slow and progressive opening of geomatics, followed by the rapid and all-encompassing opening of the Geoweb, is accompanied by a dual evolution of the "map" object. First of all, the digitization of data and the generalization of the use of GIS have led to a segmentation of the objects into layers[9].

The interoperability of both data and tools accelerates this map "dislocation" into informational artifacts that circulate independently of their initial production and contexts of use. Interoperability is a transversal notion that is at the heart of the development of the Internet, the web and the Geoweb (Haklay et al. 2008). From a single interface, it is possible to consult, visualize, map, analyze, and even download data that are located on other servers and produced by different organizations (public and private).

Interoperability opens the way to new map uses by decompartmentalizing data. Web platforms can access remote databases, and thus aggregate data from different sources. As a result of long research and negotiation work to define common standards, such as the Open Geospatial Consortium's open standards, interoperability has become a reality in the last 10 years, fully integrating itself into the daily life of cartographers and GIS experts (Percivall 2010). This new modality of dissemination and sharing of information flows, especially spatial flows, questions the borders of the spatial database, since the latter can vary according to the requests made by the user, and even adapt to the profiles.

This evolution does not suggest a disappearance of the map, quite the contrary. The second major evolution that accompanies the process of openness that was previously highlighted involves the ever-increasing place of the map in user interfaces. Most of the first GIS software put forward the programming console to process alphanumerical data, limiting the map to a "postage stamp" format. The technical origin of GIS has, in fact, profoundly characterized their design: the appropriation of GIS technologies by experts in the field of measurement (surveyors in particular) has, from the outset, oriented the ergonomics and design of software and data models. The so-called "office" GIS associated with the development of micro-computing have changed the deal. They have embedded a series of premade functionalities in more user-friendly graphic environments. Maps have taken a more central, albeit still small, place among menus and toolboxes.

---

9 This necessary structuring of information is partially in line with the techniques of traditional cartography, where separation into layers was also effective. However, the segmentation of the map into layers was based on colors, and it is now based on objects and their type.

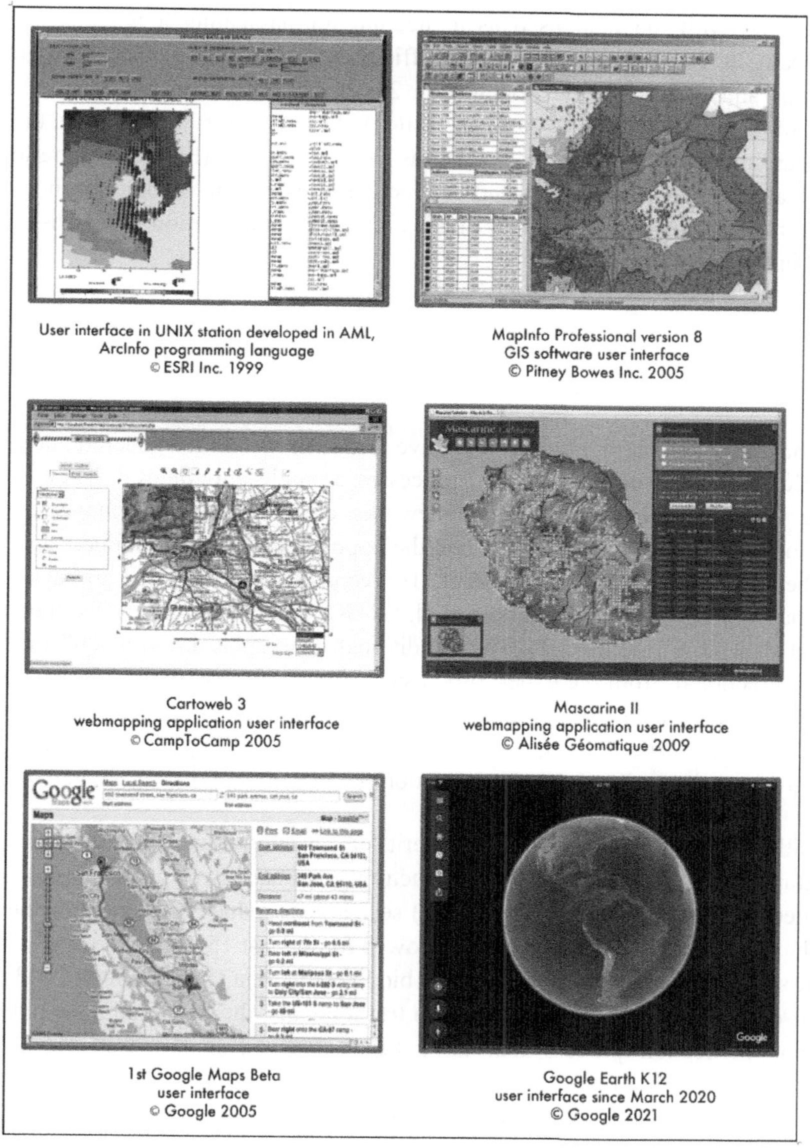

Figure 5.2. *Evolution of the map's place in the interface of geographic information systems, online sites and Geoweb applications. For a color version of this figure, see www.iste.co.uk/mericskay/cartographic.zip*

The cartographic interfaces of the Geoweb have amplified this movement by immersing the user in an almost exclusively spatial environment immediately upon connection, with the map used as a medium for navigating a universe of data (Joliveau 2011). The simplicity and efficiency of virtual globes, cartographic mashups and geoportals (de Longueville 2010) make the map the basic building block of the human–machine interfaces (HMI) of the Geoweb[10]. Over time, the map has thus improved in visibility as well as in functionality (Figure 5.2), reflecting users' appetite for its graphic and interactive appeal. This evolution shows the urgent need to continue and renew studies on the communication processes around maps in the Geoweb era.

5.3. The Geoweb era: new configurations between producers and users, amateurs and professionals

The multiple reconfigurations that have just been mentioned generate increasing complexity in the communicational processes around cartography[11]. They tend to render the traditional analytical frameworks obsolete, or even inoperative, in addition to their categorizations around the oppositions between producer and user, and between professional and amateur. To deconstruct them, let us go back to the schematics proposed by Budhathoki et al. (2008). The latter contrast two models of spatial data flow (Figure 5.3). The "traditional geographic information" model is mono-directional: from the expert producer to the expert user, and even, to a lesser extent, to the amateur user. The "voluntary geographic information" model, on the other hand, is multi-directional: users and producers as well as amateurs and experts have the capacity to disseminate information flows.

Although this diagram has the merit of visually translating the numerous geographic information flows, its structuring around two pairs of categories (amateur/professional and producer/user) seems very reductive in view of the uses that have been developing around the Geoweb for the last 15 years. We have already discussed this diagram by applying it to biodiversity data (Amelot et al. 2013). We propose to deconstruct it here in order to test it against the two evolutions linked to the communication process that have been observed in the last decade: the malleability of the map object and the individualization of data collection. Malleability and individualization lead us to rethink the categories of players that are involved in cartographic communication processes: from their creation to the uses of maps in the Geoweb era.

---

10 For more on this line of questioning, see Chapter 7.

11 For a more detailed reading on the questions surrounding the cartographic communication process, see Chapter 2.

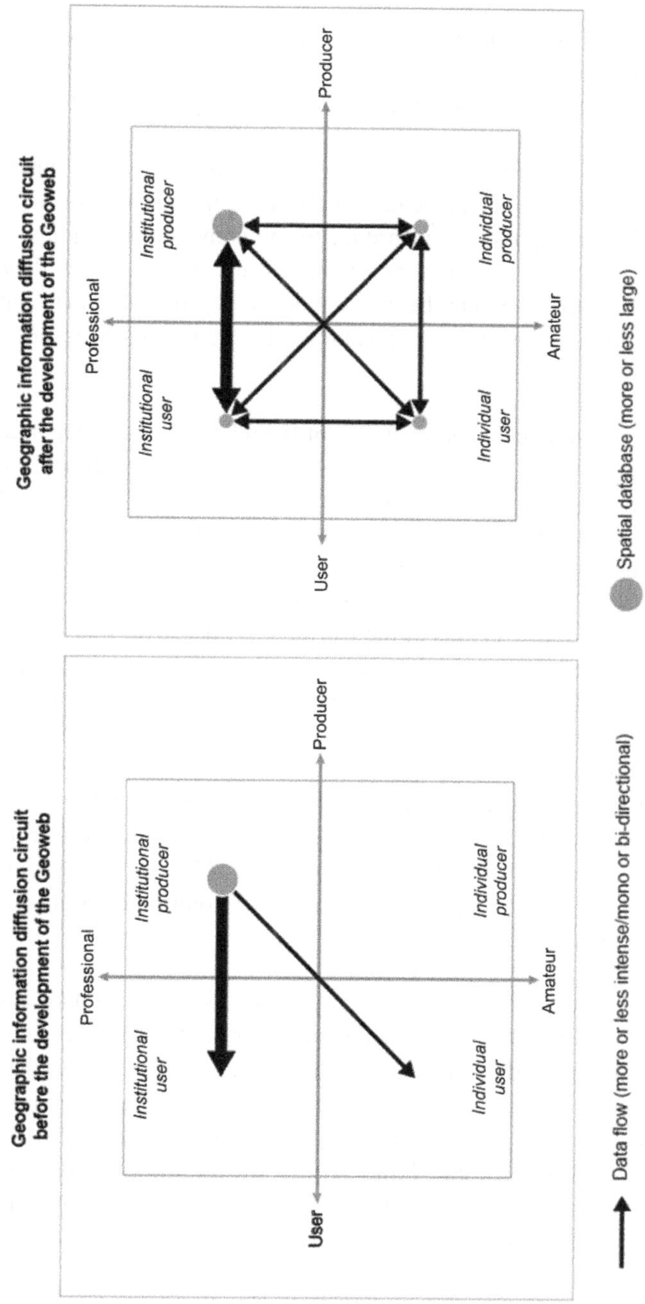

Figure 5.3. *Evolution of geographic information flow before and after the development of the Geoweb (Budhathoki et al. 2008)*

5.3.1. *The malleability of maps (and data) in the Geoweb era*

The first criticism of Budhathoki's simplifying vision is the reduction he makes in his reading of the evolution of the circulation of geographic information. Although this flow of information has indeed multiplied under the effect of the interoperability of systems, it has also produced intermediate objects that are now difficult to classify in one or other of the categories proposed by this analytical framework. The digitization of spatial data complicates the constitution of clear corpora, or even makes it impossible, according to the conventional typologies of communicational analysis that separates professionals and amateurs. As they are "malleable", in the sense that they can be easily modified by people other than the experts who created them, maps are now constituted by aggregates of information from multiple sources.

Thus, classifying geographic databases in one or the other of these categories is a challenge. The OpenStreetMap (OSM) project provides a good example of these difficulties. Often referred to as the "Wikipedia of cartography", OSM is built around multiple collaborative environments and projects that are aimed at populating and valuing a freely editable and shared global spatial database (Arsanjani et al. 2015). Created in 2004, the project has experienced strong momentum since 2007. In some regions of the world, OSM has, over time, become a reference dataset that feeds many map collections, including some institutional ones[12]. However, this collaborative system, which is open to all, has long been perceived as "amateur" cartography. This term, in its pejorative connotation, has thus served to discredit the project by highlighting the diversity of contributors and the heterogeneity of the data, which would only allow for anecdotal use. On the other hand, GIS experts and the standardized rigor of their production protocols would be the only guarantees of the credibility of "official" data. This dichotomy has been widely challenged in recent years, in terms of contributor profiles, data content and emerging uses.

At the contributor level, the OSM community looks less and less like a small group of trendsetters, disconnected from operational spheres. The annual State of the Map conferences testify to the dynamism and professionalism of the network. They are attended by both activists and professionals (institutional, academic and private companies), and the two can go hand in hand. A 2016 survey of 298 active contributors in France also highlighted the profile of contributors (Duféal and Noucher 2017). Far from constituting a monolithic block, the responses obtained do draw up a "composite profile" of OSM contributors in France. The typical profile is

12 Although not exhaustive, the census carried out on the Wiki of the French OSM gives an idea of the extent of OSM data reusability. Available at: www.openstreetmap.fr/utiliser

male (88% of respondents), almost 40 years old (38 years old on average), with a high level of qualification (58% have at least one master's degree) and working as an engineer in the private sector (31%) or for a public authority (14%)[13]. Above all, this survey shows that nearly one-third of the active participants have a degree in geomatics and integrate this activity into their professional environment. An OSM contributor in Rennes stated, "I work for IGN (National Institute of Geographic and Forest Information) during the day and contribute to OpenStreetMap at night."

The consistency of these two worlds can also be seen in the data. An analysis of the contribution practices declared in the 2016 survey shows that 19% of respondents integrate OSM into their professional environment. Moreover, the promotional discourse around the project, which tends to value the fieldwork of the "cartographer ants" constantly surveying the territory, has been largely nuanced (Noucher 2016). There are numerous modes of contribution, and a good part of them involve the integration of existing data. Several important contributors (in other words, with a sustained activity) therefore focus on identifying open sources (open data disseminated by local governments or State services), and then developing programs for their massive integration. Some 13% of the survey respondents integrate open data, but the volumes processed are often substantial and therefore represent a much larger share of the available data, as we have already been able to highlight at the scale of the Bordeaux metropolitan area (Noucher 2016).

The case of land use is exemplary from this point of view. The European CORINE Land Cover database is integrated by default, where other data are not available. These official data are then used (corrected, refined, etc.) according to the sectors and the contributors' interests. The OpenStreetMap database user is thus faced with a patchwork of data combining official fully integrated data, official data that have been integrated and reworked by a contributor, and contributory data.

Beyond the profile of contributors and the content of this cartographic database, the analysis of uses (public or private, on a local or global scale) extends this hybrid observation between amateur and professional spheres. Consequently, on a global scale, the company ESRI, a leader in GIS, announced in June 2014 that it was using OpenStreetMap data to enrich its African base maps, "thus supporting the professional uses of the OSM database"[14]. Three ready-to-use base map services (topographic background, road background, grayscale background) and three toponymy services (administrative boundaries and contours, road network, transport

13 The survey data – three sets – are available on data.gouv.fr under the ODBL license. Available at: www.data.gouv.fr/fr/organizations/projet-de-recherche-ecce-carto/#datasets

14 Press release, June 4, 2014, "On the use of OpenStreetMap data in ArcGIS Online BaseMaps", ESRI France.

network) thus exploit data from OSM for 42 African countries, where official map coverage was lacking. In the same logic, several companies specialized in online mapping such as Mapbox, Maptiler or Jawg base, a large part of their professional and paying services (base map, geocoding, route calculation, etc.) on data produced by the community.

On a local scale, local governments such as Saint-Paul-les-Dax, Digne and Montpellier (Plantin and Valentin 2013) have deployed cartographic infrastructures based on OpenStreetMap, sometimes in conjunction with other collaborative projects such as Mapillary, for panoramic photographs of public spaces. Flora Hayat's dissertation arguably provides one of the most successful examples of integrating OpenStreetMap into a cartographic publishing process (Hayat 2019). Her immersion in Michelin's cartography department allowed her to assess the conditions of integration of collaborative data such as OSM in a large-scale cartographic production chain. The production of the Michelin map of Guadeloupe, Saint-Martin and Saint-Barthélemy (reference 137 zoom overseas) at the 1/80,000 scale is the result (Figure 5.4) and demonstrates the fusion between the professional sphere and the amateur community.

Figure 5.4. *Title block and box of Michelin map No. 137 on Guadeloupe at 1/80,000, integrating the mention of OSM contributors (2018 edition). For a color version of this figure, see www.iste.co.uk/mericskay/cartographic.zip*

The hybridization of amateur and professional practices is becoming increasingly significant in online usage well beyond the field of cartography. Flichy's (2010) book *Le Sacre de l'amateur : sociologie des passions ordinaires à l'ère numérique*, and Cardon's (2010) *La Démocratie Internet : promesses et limites*, published the same year, highlight the symbolic weakening of the separation between amateurs and professionals, in connection with the development of digital culture. The term *pro-am* (a portmanteau of professional/amateur) was coined to emphasize the importance of knowledge and skills acquired through experience, particularly within virtual communities. The intermingling of the positions of producer and user is exemplary in the field of cartography, as in the OpenStreetMap project or Geonames project for toponymy. These free and open databases are more and more integrated in the cartographic interfaces that are developing on the Geoweb, whether they are public or private, professional or associative. In this sense, they are so present in online mapping services that we sometimes forget to query them. OpenStreetMap has thus become a base map in just a few years, but also, above all, an essential data producer, whose global coverage and quality of data deserve to be contextualized and questioned, as the disparities between countries or regions of the world are high (Verburg et al. 2011). Geonames, meanwhile, serves as the basis for geocoding many web mapping applications. Its heterogeneous coverage, reflecting the existing digital divides on a planetary scale (Graham and De Sabbata 2015), deserves particular attention. A fine critical analysis of these new reference systems, beyond the incantatory speeches, appears to be essential. Such work contributes to rethinking the categories of players proposed by Budhatoki, as well as to considering their hybrid nature in light of current dynamics (Figure 5.5).

The study of contributors (profiles, practices and motivations) and data (production, combinations and uses) suggests avoiding the simplifying temptation to divide institutional geographic information produced by and for public authorities, on the one hand, and voluntary geographic information produced by amateurs, on the other hand. Although the two approaches remain and do not fully combine with one another, they coexist, reconfigure and ultimately tend to hybridize. As Denis and Marquet point out in a study on the collaboration between the OSM community and the National society of French railroads (SNCF):

> Being attentive to these shifts and hybridizations also prevents us from making a clear separation between cartographic initiatives that involve amateur contributors, on the one hand, and those that involve professionals or experts, on the other. (Denis and Marquet 2019, p. 66)

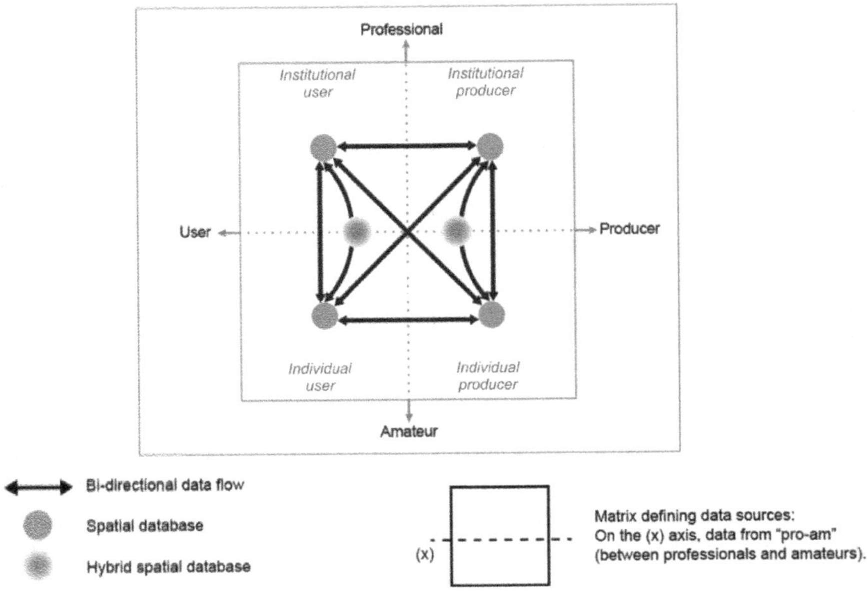

Figure 5.5. *Complexity and hybridization of geographic information flow (1/2) (according to Budhathoki et al. 2008)*

By making the *pro-am* figure emerge, it is then a question of redefining the amateur, not as the antonym of a professional, with negative connotations of dilettantism and incompetence, but as "the one who loves" and can also be led to invest in cartographic projects, so as to professionalize a passion.

5.3.2. *Individualization of sources and egocartography*

Another effect of the digital transition is the increasing amalgamation between two previously relatively sealed worlds: public and private space. In terms of cartography, this change is reflected in the individualization of map production. Several converging developments are starting to personalize maps more and more around users.

The development of volunteered geographic information (VGI) has led to the increasing externalization of the production of geographic content by voluntary citizens, who have become "sensors" of their environment with, for example, the development of platforms for publishing and sharing georeferenced photographs. The citizen "sensor" of his environment can also become a citizen that is "recorded" by his own activities using Geoweb tools. Thus, the movement of the quantified self, or self-measurement, which brings together tools, principles and methods that allow each person to measure his or her personal data (activities on social networks, sports practices, etc.), analyze them and share them, can bear witness to this individualization of data (Gadenne 2012). Personal quantification is indeed a perfect illustration of the (technical) forms of contemporary individualism, where the generalization of the autonomy standard is accompanied by the injunction to formulate and share a self-figure (Pharabod et al. 2013).

Although the users of these systems are voluntarily engaged in the production of (largely geographic) data, several authors (Fisher 2012; Caquard 2013) have hijacked this notion to highlight that, beyond this contributory geography, involuntary geographic data sensor devices were also being deployed, and proposed the notion of inVolunteered Geographic Information (iVGI), like the exploitation of digital traces left by borrowing self-service bicycles, publishing a tweet or using a mobile navigation application. The citizen sensor thus also becomes a captured citizen, who generates information (often spatialized) about their environment or activities. These data, collected in the form of digital traces, which Beaude (2015) emphasizes are “eminently spatial”, can then be remobilized by various services used to contextualize and personalize the maps made available for consultation.

This individualized and near-automated production of (personal) data is not neutral in the cartographic communication process, because the new data produced feed back into the services offered to Geoweb platform users. While e-commerce companies such as Amazon have long used these suggestions based on algorithmic predictions, the emergence of web personalization has also spread to cartography. Since 2013, Google Maps has relied on a contextualization algorithm that aims to personalize the map by integrating the traces left on the Internet by the user (search history on the Google engine, geographical position, email content, messages on social networks, etc.). Beyond the display of markers responding to a search, it is the map background itself that is affected by this personalization. The appearance of points of interest (shops, restaurants, train stations, etc.), the hierarchy of routes, the symbology of objects and the toponymy evolve as the search area becomes more precise and depends on the user (his profile as well as his positioning). Individual expressions of spatial interests then take precedence over shared reference systems (Joliveau 2013).

This personalization extends to navigation assistance by proposing travel scenarios based on routes that have already been taken. Thus, while traditional cartography produces selectivity to generate a common medium for discussion, Big Data is interested in singular profiles and generates algorithmic bubbles that personalize maps (Kliman-Silver et al. 2015). In turn, the individual datacaptured guides the content offered to each Internet user (Ertzscheid 2013).

With the development of geolocation systems, users are at the heart of cartographic interfaces, adapting to their position (to place them in the center of the map), their direction (by adjusting the orientation of the map) and moving with them (the map is mobile and moves as they travel). On the Geoweb, users participate in the production of cartographic information while consuming it (Quesnot 2021). Web mapping platforms, especially in their mobile versions, place the user at the heart of their strategy, by proposing and adapting a map that is developed around them and from them, which is a form of egocartography. In his study of Google Maps, Ström (2020) speaks of "automated and atomizing processes" to evoke this algorithmic personalization that places us, alone, at the center of the world.

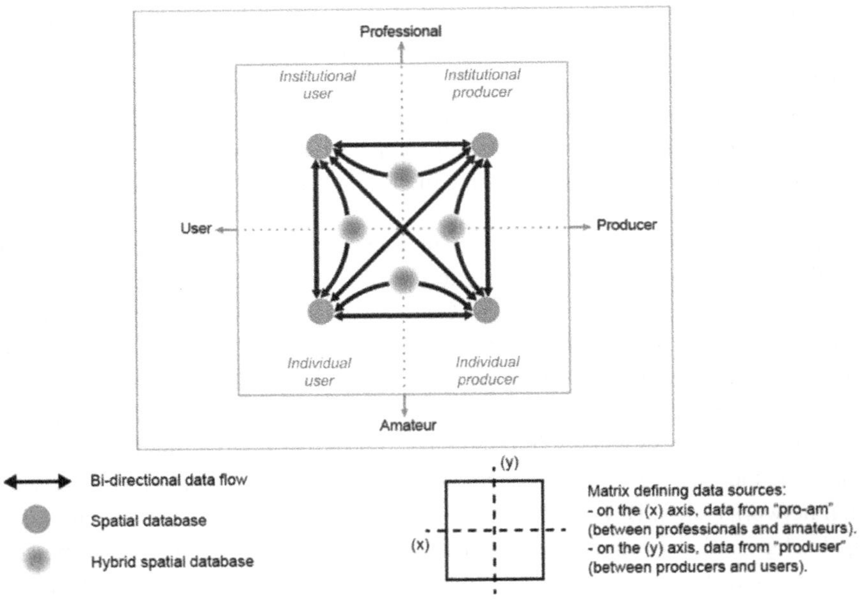

Figure 5.6. *Complexity and hybridization of geographic information flow (2/2) (according to Budhathoki et al. 2008)*

This hyperindividualization of maps makes the traditional distinction between the producer and user delicate or even obsolete, since the user now fully participates (but also surreptitiously) in its construction. The word *produser* translates this blurring of the line between the notion of production and the notion of use of information (Figure 5.6). Popularized by Bruns (2008), the notion of *produsage* characterizes collaborative encyclopedias such as Wikipedia, as well as photo-sharing platforms such as Instagram or Flickr or citizen journalism sites such as Indymedia or OhmyNews. While in the past, cartography was generally produced within a professional framework, by a group of experts with a monopoly, and then circulated without being altered too much, today this framework is exploding, and the circulation and use of data are becoming part of the very production of information.

5.4. Conclusion

Communication studies have largely contributed to rethinking the map not as a simple tool for spatial representations but as a mediation with its own characteristics, taking place between transmission and reception. The intersection between information theory and cartography has led to a significant amount of standardization work, aimed at increasing efficiency in the transmission of information, as shown by the work of Bertin (1967). However, the diversification and complexity of both the production and the uses of maps leaves a certain number of categories and analytical frameworks inactive, at least partially.

The digital transition that has been accelerating since the end of the 1990s, with the emergence of the Geoweb, has led to interwoven difficulties that complicate the analysis of map communication. Faced with combined issues, like that of a "Russian doll" (Gautreau and Noucher 2022), it is now particularly complex to grasp the ins and outs of the communicational process of maps. If the objects are difficult to grasp, and the players are numerous and vague, how can we succeed in understanding and formalizing the way in which the map circulates and is mobilized? Moreover, favored by the Internet and the networking of information, the map object is now much more difficult to define than it used to be, because it is composed of multiple digital artifacts that players can endlessly aggregate or disaggregate, whether they are experts in cartography or not. Under these conditions, identifying who does what in contemporary mapping raises formidable methodological questions. This is why the malleability and individualization of spatial data must lead us to consider a mix in conventional typologies, and to think about the hybridization of the resulting cartographic products. However, although the binary oppositions between producer and user, or between professional and amateur, are becoming blurred, communication studies (CS) remain essential disciplinary allies to studies relating to the map and, in particular, to critical cartography.

Thus, in a thematic issue of the *Communication & Langages* journal, coordinated by Robert and Souchier in 2008, the contributions of the heterogeneity of theoretical and methodological approaches in CS are underlined in order to study the new semiotic properties of digital maps on the Internet. According to Plantin (2014), CS also emphasizes the lifecycle of data: from prior processing to its mapping and the various mobilizations of the map. To shift the gaze from the great cartographic narratives, which for a long time remained focused on the production of national institutes, we propose studying the "small maps of the web" (Noucher 2017). This expression specifically refers to the work in communication studies of Candel, Jeanne-Perrier and Souchier on the "small forms of the web" (Candel et al. 2012), new forms of writing that have been emerging on the Internet for the last 10 years. Composite applications (mashup) are a good example: this writing technique is characteristic of the Internet; it offers the possibility of aggregating content from various sites in order to create a new site. Among all the possible types of mashups (textual, iconographic, musical), cartographic mashups are the most frequent, in particular thanks to the success and the ease of reuse of platforms such as OpenStreetMap or Google Maps, to name but two. The small maps of the web that exploit these new possibilities thus participate in a renewal of the forms of cartographic writing, and are now fully part of the web editorial writing toolbox. It is thus advisable not to analyze them as isolated compositions, but as elements of an editorial grammar whose framework of enunciation exceeds the map alone, a fact that obliges critical cartography to consider all the media interrelations of these maps and not to consider them in isolation.

As a result, Koláčný's model of communication and the four postulates mentioned in the introduction, which isolate the map, become partially obsolete[15]; similarly, Budhathoki's categorizations seem partially obsolete. Taking a serious approach at the thousands of small maps on the Geoweb produced by activists, journalists, political actors, artists, hackers, etc., makes us think of the communicative processes of cartography in a different way. As Dodge and Perkins (2015) point out, it is undoubtedly more complex to deconstruct current cartographic devices, where diverse spheres of players and multiple technologies are intertwined, than the devices of yesteryear, where cartographic power was held by a few experts. Integrating an enlarged vision of the map object that includes all the informational artifacts associated with it (from virtual globes to GPS tracks, from spatial databases to geolocation applications, etc.), as well as the diversity of players that gravitate around it (digital giants, institutions, associations, individuals, etc.), allows us to glimpse abysmal but stimulating prospects in order to rethink critical approaches to cartographic communication.

15 Chapter 2 presents the same conclusion by approaching this problem through the proposal of new models of cartographic communication.

5.5. References

Amelot, X., Couderchet, L., Noucher, M. (2017). Données institutionnelles et données contributives sur la biodiversité, quelle légitimité ? *Mappemonde*, 120.

Arsanjani, J.J., Zipf, A., Mooney, P., Helbich, M. (2015). *OpenStreetMap in GIScience. Experiences, Research, and Applications*. Springer International Publishing, New York.

Beaude, B. (2015). Spatialités algorithmiques. In *Territoires et traces numériques*, Severo, M., Romele, A. (eds). Presses des Mines, Paris.

Bédard, Y. (1993). Explosion informationnelle et révolution technologique : la naissance d'un nouveau défi. In *Séminaire sur les SIT de l'École Polytechnique Fédérale de Lausanne*.

Bertin, J. (1967). *Sémiologie graphique. Les diagrammes, les réseaux, les cartes*. Gauthier-Villars-Mouton, Paris.

Bougnoux, D. (1991). *La communication par la bande*. La Découverte, Paris.

Budhathoki, N.R., Bruce, B., Nedovic-Budic, Z. (2008). Reconceptualizing the role of the user of spatial data infrastructure. *GeoJournal: An International Journal on Geography*, 72(3), 149–160.

Bruns, A. (2008). *Blogs, Wikipedia, Second Life, and Beyond: From Production to Produsage*. Peter Lang, New York.

Candel, E., Jeanne-Perrier, V., Souchier, E. (2012). Petites formes, grands desseins. D'une grammaire des énoncés éditoriaux à la standardisation des écritures. In *L'économie des écritures sur le Web*, Davallon, J. (ed.). Hermès-Lavoisier, Cachan.

Caquard, S. (2013). Collective cartographies in the social media era. *Progress in Human Geography*, 38(1), 141–150.

Cardon, D. (2010). *La Démocratie Internet. Promesses et Limites*. Le Seuil, Paris.

Cardon, D. (2015). *À quoi rêvent les algorithmes, nos vies à l'heure des big data*. Le Seuil, Paris.

Crall, A.W., Newman, G.J., Stohlgren, T.J., Holfelder, K.A., Graham, J., Waller, D.M. (2011). Assessing citizen science data quality: an invasive species case study: Assessing citizen science data quality. *Conservation Letter*, 4(6), 433–442.

Crampton, J. (2001). Maps as social constructions: Power, communication and visualization. *Progress in Human Geography*, 25(2), 235–252.

De Longueville, B. (2010). Community-based geoportals: The next generation? Concepts and methods for the geospatial Web 2.0. *Computers, Environment and Urban Systems*, 34(4), 299–308.

Debarbieux, B. and Hirt, I. (2022). *The Politics of Mapping*. ISTE Ltd, London, and John Wiley & Sons, New York.

Denis, J. and Marquet, C. (2019). De la production à la maintenance : faire exister collectivement des données géographiques ouvertes. *Questions de communication*, 36, 63–83.

Desbois, H. (2015). La carte et le territoire à l'ère numérique. *Socio, la nouvelle revue des sciences sociales*, 4, 39–60.

Dodge, M. and Perkins, C. (2015). Reflecting on J.B. Harley's influence and what he missed in "deconstructing the map". *Cartographica. The International Journal for Geographic Information and Geovisualization*, 50(1), 37–40.

Duféal, M. and Noucher, M. (2017). Des TIC au TOC. Contribuer à OpenStreetMap : entre commun numérique et utopie cartographique. *Netcom*, 31(1/2), 77–98.

Ertzscheid, O. (2013). Le Web : entre stratégies prédictives et technologies injonctives. In *Séminaire Identités numériques*. Institut des Sciences de la Communication du CNRS, Paris.

Feyt, G. (2004). The space and place trades versus geographic information technologies: Babel and Esperanto. In *Organizational Facets of GIS*, Roche, S. and Caron, C. (eds). ISTE Ltd, London, and John Wiley & Sons, New York.

Flichy, P. (2010). *Le sacre de l'amateur. Sociologie des passions ordinaires à l'ère numérique*. Le Seuil, Paris.

Fischer, F. (2012). VGI as big data: A new but delicate geographic data-source. *GeoInformatics*, 3, 46–47.

Gadenne, E. (2012). *Le Guide Pratique du Quantified Self*. Fyp, Limoges.

Gautreau, P. and Noucher, M. (2022). Farewell to maps: Reformulating critical cartography in the digital age. In *The Politics of Mapping*, Debarbieux, B. and Hirt I. (eds). ISTE Ltd, London, and John Wiley & Sons, New York.

Goodchild, M.F. (2007). Citizens as sensors; the world of volounteered geography. *GeoJournal*, 69(4), 211–221.

Graham, M. and De Sabbata, S. (2015). Mapping information wealth and poverty: The geography of gazetteers. *Environment and Planning A: Economy and Space*, 47(6), 1254–1264.

Haklay, M., Singleton, A., Parker, C. (2008). Web mapping 2.0: The neogeography of the GeoWeb. *Geography Compass*, 2(6), 2011–2039.

Harley, J.B. (1989). Deconstructing the map. *Cartographica. The International Journal for Geographic Information and Geovisualization*, 26(2), 1–20.

Hayat, F. (2019). Production des biens communs numériques et usages cartographiques. PhD Thesis, University of Paris, Paris.

Jacob, C. (1992). *L'Empire des cartes. Approche théorique de la cartographie à travers l'histoire*. Albin Michel, Paris.

Joliveau, T. (2010). La géographie et la géomatique au crible de la néogéographie. *Tracés*, 3(10), 227–239.

Joliveau, T. (2011). Le géoweb, un nouveau défi pour les bases de données géographiques. *L'Espace géographique*, 2(2), 154–163.

Joliveau, T. (2013). Chacun sa carte ? Le nouveau Google Maps. *Mappemonde*, 110.

Joliveau, T. (2020). Une révolution numérique de la géographie ? Le cas de la géomatique. *Histoire de la recherche contemporaine*, 9(1), 21–34.

Joliveau, T., Noucher, M., Roche, S. (2013). La cartographie 2.0, vers une approche critique d'un nouveau régime cartographique. *L'Information géographique*, 4(4), 29–46.

Joliveau, T., Noucher, M., Couderchet, L., Caquard, S. (2017). Enseigner le géoweb par la pratique et la critique. Retour sur sept années de cours à distance. *Ingénierie des Systèmes d'Information*, 22(5), 11–33.

Kitchin, R., Lauriault, T., Wilson, M. (2017). *Understanding New Spatial Media*. SAGE, New York.

Kliman-Silver, C., Hannak, A., Lazer, D., Wilson, C., Mislove, A. (2015). Location, location, location: The impact of geolocation on web search personalization. In *Proceedings of the 2015 Internet Measurement Conference (IMC '15)*. Association for Computing Machinery, New York.

Koláčný, A. (1969). Cartographic information, a fundamental concept and term in modern cartography. *The Cartographic Journal*, 6(1), 47–49.

Miller, C. (2006). A beast in the field: The Google Maps mashup as GIS/2. *Cartographica. The International Journal for Geographic Information and Geovisualization*, 41(3), 187–199.

Montello, D.R. (2002). Cognitive map-design research in the twentieth century: Theoretical and empirical approaches. *Cartography and Geographic Information Science*, 29(2), 283–304.

Noucher, M. (2015). À Bordeaux, les cartes font le pont. *Mappemonde*, 115.

Noucher, M. (2017). *Les petites cartes du Web. Approches critiques des nouvelles fabriques cartographiques*. Presses de l'École normale supérieure, Paris.

Noucher, M. (2019). Information géographique volontaire. *Encyclopédie en libre-accès Hypergeo* [Online]. Available at: https://www.hypergeo.eu/spip.php?article725/ [Accessed 8 July 2021].

Percivall, G. (2010). The application of open standards to enhance the interoperability of geoscience information. *International Journal of Digital Earth*, 3(1), 14–30.

Pharabod, A.S., Nikolski, V., Granjon, F. (2013). La mise en chiffres de soi. Une approche compréhensive des mesures personnelles. *Réseaux*, 177(1), 99–129.

Plantin, J. (2014). L'avènement de la carte comme médiation. *Questions de communication*, 25(1), 309–326.

Plantin, J. and Valentin, J. (2013). Données ouvertes et cartographie libre : autour du cas de Montpellier. *Les Cahiers du numérique*, 1(1), 85–107.

Pornon, H., Nedovic-Budic, Z., Pinto, J. (2004). Geographic information and interorganizational partnerships. In *Organizational Facets of GIS*, Roche, S. and Caron, C. (eds). ISTE Ltd, London, and John Wiley & Sons, New York.

Purves, R., Edwardes, A., Wood, J. (2011). Describing place through user generated content. *First Monday*, 16(9) [Online]. Available at: https://firstmonday.org/ojs/index.php/fm/issue/view/686 [Accessed 15 July 2021].

Quesnot, T. (2021). Décrypter l'innovation de la cartographie numérique grand public : entre marchandisation, appropriation et représentation égographique de l'information. *Mappemonde*, 131. [Online]. Available at: https://journals.openedition.org/mappemonde/5728 [Accessed on 15 July 2021].

Robert, P. and Souchier, E. (2008). La carte, un média entre sémiotique et politique. *Communication & langages*, 158, 25–106.

Roche, S. (2000). *Les enjeux sociaux des systèmes d'information géographique*. L'Harmatan, Paris.

Shannon, C.E. (1948). A mathematical theory of communication. *Bell System Technical Journal*, 27(3), 379–423.

Ström, T.E. (2020). Journey to the centre of the world: Google Maps and the abstraction of cybernetic capitalism. *Cultural Geographies*, 27(4), 561–579.

Sui, D., Elwood, S., Goodchild, M. (eds) (2013). *Crowdsourcing Geographic Knowledge. Volunteered Geographic Information in Theory and Practice*. Springer, Berlin.

Turner, A. (2006). *Introduction to Neogeography*. O'Reilly Media, Newton.

Verburg, P.H., Neumann, K., Nol, L. (2011). Challenges in using land use and land cover data for global change studies. *Global Change Biology*, 17(2), 974–989.

Wood, D. (1992). *The Power of Maps*. Guilford Press, New York.

6

Rethinking Cartography on the Geoweb: Principles, Tools and Modes of Representation

Boris MERICSKAY
ESO, University of Rennes 2, France

6.1. Introduction

Cartography's conversion to the digital world, which has been going on for more than thirty years, has been largely disrupted by the advent of the web. As a result of GIS linking with Internet technologies and geolocation systems, a particular form of the web has emerged, the Geoweb, which characterizes the networking of databases, applications and users of geographic information (Haklay and Singleton 2008; Scharl and Tochtermann 2009). With the Geoweb, cartography has profoundly evolved in terms of techniques, the public, uses and modes of representation. Anchored in the technological context and practices of the social web, this geographic web allows the general public to read and draw maps using user-friendly tools that are increasingly similar to those of specialists (Mericskay and Roche 2011).

In this era of Open Data, Big Data and real-time data, the traditional logics of use, representation and visualization of geographic information are changing, giving rise to the emergence of a new cartographic language (Elwood 2011; Muehlenhaus 2013). In order to meet the challenges of (geo)visualizing increasingly massive and complex data (volume, readability, graphical rendering, interaction, etc.), web mapping solutions are gradually moving away from the world of (thematic)

Cartographic Communication,
coordinated by Boris MERICSKAY. © ISTE Ltd 2023.

cartography and GIS, and instead toward that of data exploration and visualization (Mericskay 2016). This new cartographic regime (Joliveau et al. 2013) questions cartographers, insofar that maps, which are omnipresent, change both their appearance and function in the daily lives of individuals and organizations.

Although technological innovations are the main driving force behind this form of cartography, questions that come up from new cartographers must be taken into account and put into perspective with regard to the current uses of web maps. The modes of cartographic representation that are developing and becoming widespread on the Geoweb must also be questioned, be it in terms of design methods, the types of interactions they propose or their efficiency. This is this chapter's objective, which proposes a synthesis of the modalities and trends of cartography on the Geoweb, by questioning their characteristics and the underlying challenges in terms of cartographic communication.

In order to understand cartography on the Geoweb and how it transforms the logic of geographical data representation, four perspectives of analysis will be developed in this chapter. First, we will present the sociotechnical foundations of the Geoweb, which condition the new cartographic practices. We will then address the constituent elements of web maps, namely base maps and the forms of interaction between users, maps and data. We will then present a synthesis of the cartographic language of the Geoweb, exploring the main visualization methods of spatial big data (markers, dot maps, heatmaps, marker clustering and 3D maps). Lastly, we will come back to the question of instrumentation through an overview of current tools and a reflection on technology's place in the mapping process on the web.

6.2. Coming back to cartography in the Geoweb

6.2.1. *The Geoweb, an environment for reading and writing maps*

With the Geoweb, cartography has taken its place within networks and information flows (Hudson Smith et al. 2009; Plantin 2014). Through its new properties of integration, sharing and blending, the web map behaves as an online communication tool, and an aggregator of digital information and content. Unlike the web mapping of the early 2000s, where the map was an instrument for representing geographical data, on the Geoweb, the map is mostly a tool for spatial representation of data, insofar as it is the web as a whole that takes on a geographical dimension. On the majority of web pages (news, social networks, businesses, photos, etc.), we find information related to location in a direct or indirect way, such as geographical coordinates, addresses or just place names (Scharl and Tochtermann 2007). The geographical dimension of web content through the location component becomes metadata, an additional piece of information that is added to and enriches existing content. In this dynamic of spatially indexing digital content, the map is emerging as an effective medium that, in addition to the possibility of visualizing information, allows its spatialization (Joliveau et al. 2013). On the Geoweb, it is no longer a question of querying databases to make maps, but of using cartographic interfaces to search and navigate the entire Internet (Joliveau 2011).

Beyond the spatial visualization of heterogeneous content, many Geoweb applications also offer a set of tools to create and represent geographical data. On the one hand, maps are becoming more and more contributive and participative. With the development of crowdsourcing and volunteered geographic information (Sui et al. 2012), the general public is positioning itself as a full-fledged player in creating and updating institutional, private, and community spatial databases (Mericskay and Roche 2011). On the other hand, online instrumentation allows the production of maps, of varying degrees of complexity, to be disseminated and shared on the web. Whether it is a matter of placing pins on a base map, drawing areas or mapping statistical data, Geoweb tools have largely integrated the functions of professional CAD and GIS software (Palsky 2013). After long being just a matter for specialists, cartography is now becoming more democratic in a variety of ways, profoundly changing the way we see and represent the world.

6.2.2. *Cartographic communication in the movement of data (geo)visualization*

In a context of proliferation, openness and mixing of geographical content, the visualization and exploration of spatial data in dynamic and interactive forms appear as full-fledged steps in the analytical work, as well as in the dissemination of data, especially geographical data. The question of geovisualization, popularized in the 1990s in the movement of scientific visualization and exploratory data analysis (MacEachren and Kraak 1997, 2001), truly finds its meaning today with the Geoweb and Big Data.

Geovisualization[1] as a set of methods and tools for exploring, analyzing, synthesizing and visually presenting large spatial data sets is a particularly dynamic focus in the development of online mapping. Unlike GIS, the focus is not on data analysis and communication, but rather on visualizing and exploring data to facilitate understanding (Kraak and Ormeling 2010; Zastrow 2015). It is more of an "exploratory" web mapping that stimulates thinking in an intuitive way. In this

1 For a more complete definition of geovisualization, see Chapter 7.

context, the map is not fixed; it is built according to the available data sets, the functionalities of the interfaces as well as the profiles and objectives of the users in a dynamic and interactive approach.

The dynamic aspect implies the possibility of moving around the map and changing the zoom level. By changing the scale of visualization, the data adapt to the scalar level, both in terms of formatting and the nature of the data. The interactive aspect implies the possibility of adding data, overlaying them, changing the base map and, especially, interacting through contextual windows or selection, sorting or filtering tools. Faced with these new forms of interaction between the user, the map and the data, the role of the user in the visualization strategies is changing. The digital map is no longer really on the side of the designer, but rather on the side of the user who builds it progressively (MacEachren and Kraak 2001). The distinction between map users and map designers is no longer fixed, and both approaches and methods of research and assessment around cartography are transforming (Griffin et al. 2017). The Geoweb, inseparable from the operating modes of the web, is profoundly renewing the properties of the map as a mediation by shifting the traditional separation between sender and receiver (Plantin 2014)[2].

Interacting with the data, manipulating them and changing the scale all offer new possibilities in terms of exploring and investigating the mass of spatialized information available on the Geoweb. The map has become a heuristic tool, making it possible to access details that are invisible at first sight in the mass of processed data. This exploratory mode shifts the methodological guidelines from an explanatory logic to a comprehensive logic of phenomena. Making data "visible" is not enough to make it understandable. The reader must also be able to interpret and understand what the data refer to, both in terms of the information represented and the graphic forms used. In this sense, the map is only one element among others of online data exploration and visualization interfaces.

One of the major trends in cartographic applications on the Geoweb is to go beyond a simple map to propose dashboards. These interfaces that centralize and combine several modes of data visualization, such as graphics (histograms, scatterplots, pie charts, etc.), lists, key figures and maps, are becoming increasingly important on the Geoweb and in society in general (Kitchin et al. 2015). Originating from the field of business intelligence, the purpose of a dashboard is to provide the user with a summarized display of large amounts of formatted information in order to quickly make sense of the data it presents (Few 2006). Applied to the field of geo-information, a spatial dashboard is a tool that presents simple, explicit and structured views of key indicators that can be useful to users (Figure 6.1).

2 For more information on this topic, see Chapters 2 and 5.

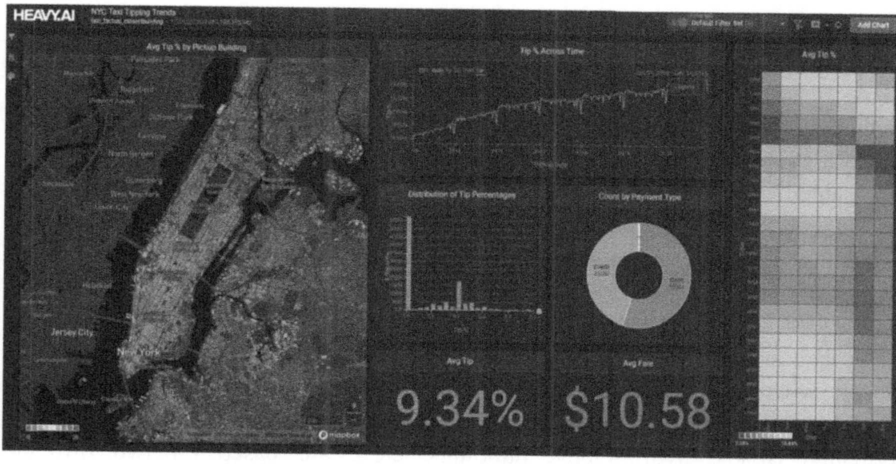

Figure 6.1. *OmniSci dashboard interface to explore 1.3 billion cab rides in New York City (source: OmniSci). For a color version of this figure, see www.iste.co.uk/mericskay/cartographic.zip*

These applications refer to Schneiderman's (1996) mantra of information visualization, which revolves around a first overview, changes in scale and filters (zoom and filter) and more details on demand. Beyond simply representing and synthesizing a large amount of information in various forms, by combining maps and graphs, dashboards also offer the possibility to explore data in other ways. A zoom on the map can, for example, update the graphs by adapting the represented sample to the scope of the map. And, vice versa, a selection on a chart or a list will update the data displayed on the map and the key figures.

6.3. Base maps and interactivity: a look at the basic elements of Geoweb maps

Once the main cartographic principles on the Geoweb have been established, the second part's objective is to return to two elements of online maps in more detail. Here, we propose addressing the question of base maps as referential and that of interactivity between user, map and data at the center of important issues in the construction of web maps, which play a role in communication.

6.3.1. *Base maps as referential of web maps*

As central elements of cartographic mashups[3], base maps constitute the mediation supports on which (spatial) data are superimposed and organized (Batty et al. 2010). As repositories, they allow maps to be “dressed up” and enrich the represented data with referential information (Raposo et al. 2020). Proposed by public, private or community players, base maps are increasingly accessible, numerous, accurate, up-to-date and, above all, diversified in terms of styles (Figure 6.2).

From a technical point of view, base map services are based on the principle of a mosaic, where the world is divided into map tiles[4], whose content adapts to the level of scale of visualization. These tile pyramids, which offer data sets that are already formatted (roads, hydrography, buildings, etc.), are based on two different technologies: raster tiles (based on images) and vector tiles (based on vector data). Unlike GIS software, where the user can use a wide range of projection systems, on the Geoweb, the applications are essentially based on the Web Mercator projection (EPSG: 3857). Widely popularized by Google Maps, this variation of the Mercator projection, which was designed to harmonize and optimize the diffusion of maps on the web, is used by the vast majority of public, private and community map providers (Eckert and Jégou 2008; Battersby et al. 2014). Some solutions, however, offer the use of other projection systems, but in rather marginal ways.

The use of vector tiles compared to raster tiles has many advantages, both in terms of performance (transfer speed), rendering capabilities (fluidity) and style customization (Antoniou et al. 2009; Gaffuri 2012; Netek et al. 2020). It is thus possible to quickly create one’s own base map style according to taste, the theme treated or a graphic charter. This is particularly interesting, as the design of base maps in terms of style and content is increasingly centered around users (Beukelaar 2018). This new dynamic, to a large extent, relies on OpenStreetMap’s open source data, like the base maps produced in the early 2010s by companies like Stamen, Mapzen or Mapbox, which launched a new way of viewing and using these base maps by introducing a design aspect.

3 A mashup, or composite application, consists of combining content from various sources behind a unified graphical interface to offer a hybrid service, such as geographic data from one site on the base map of another site.

4 A map tile is a small area, adapted to the zoom level, containing a piece of the base map (as image or vector data).

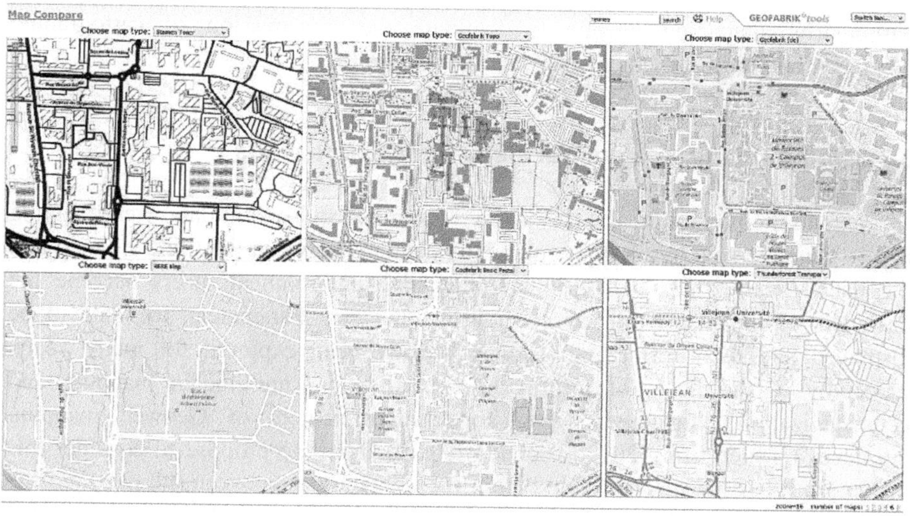

Figure 6.2. *Examples of base map styles based on OSM data (sources: Geofabrik, OpenStreetMap). For a color version of this figure, see www.iste.co.uk/mericskay/cartographic.zip*

From minimalist to vintage, monochrome or artistic styles, thousands of pre-configured base maps are available on platforms like Snazzy Maps, Mapbox Gallery or Open Map Styles. It is also possible to edit your own base map styles by modifying the data present, their formatting properties or the labels. There are numerous online toolboxes to easily and quickly customize them such as Mapbox Studio, Vector Tile style Editor from ESRI, Map Style from Google, or Map Style Sheet Editor from Microsoft. Once designed, these customized base maps can be used online or within GIS software, thus opening the way to wider uses.

The challenges surrounding the customization of base map styles refer to questions about the data to be displayed and their formatting characteristics (color, opacity and typography). This setting, which is the result of decisions made upstream during the generation of tiles, is an important step in the design of online maps (Beukelaar 2018; Panchaud and Hurni 2018). For example, it is interesting to observe that with this new generation of tools, which allow infinite customization of base maps with a few clicks, we are paradoxically witnessing a certain standardization of styles available on the Geoweb. The "plan" style, popularized since 2005 by Google Maps, has thus become a standard, widely imitated by the majority of base map providers and typically expected by users (Dubrava 2017). The question of stylization, as well as the expressiveness of these base maps, is another particularly interesting avenue to facilitate the transmission of cartographic messages, and ultimately make maps more effective[5].

6.3.2. *Interactivity as a central element of online maps*

Beyond the possibilities of displaying base maps and spatial data, web mapping is also characterized by multiple interactive features between the user, the map and the data[6] (Roth 2012, 2013). In an interactive environment, users become the designers of the maps, quickly configuring the view parameters, the data to be displayed and their formatting. Having already been explored for many years, especially in the fields of information visualization, multimedia mapping and geovisualization, interactivity is a constitutive element of Geoweb mapping interfaces (Lienert et al. 2012; Tocholko 2016). Here, we propose to consider this matter from two perspectives, first from the interactivity between the user and the map view, and second from the forms of interactions between the user, the display and the representation of spatial data (Table 6.1).

Map view	Display and representation
Pan	Display of layers (menu)
Zoom	Data filtering and sorting
Inclination	Search bar
Rotation	Time bar
Base maps	Integration of external data
Geographic tabs	Annotation/drawing
Geolocation	Calculation/measurement/analysis
Geocoder	Brushing
Swipe maps	Contextual information
Synchronized maps	Adaptive display
Minimaps	Adaptive symbology

Table 6.1. *Summary of the main forms of interaction between users, maps and data on the Geoweb*

First of all, moving (pan) and changing the visual scale (zoom) is a fundamental part of online mapping. Popularized by consumer applications such as Google Maps,

5 To further develop this topic of styles and customization of base maps, see Chapter 7.

6 For more on interactivity in cartography, see Chapter 7.

these actions are now widely known and the ability to move a map and zoom, either with a mouse or via finger movements on a phone, has become a typical everyday act in our lives. With the development of WebGL, which allows for three-dimensional spatial representations, it is possible to pitch and rotate maps to get away from the traditional north-facing top view. Changing base maps on the fly, with opacity or hybridization options, has also become commonplace within mapping interfaces.

In terms of view centering, the user can adjust the map according to his/her location or by entering an address in a search bar (geocoding). For a more focused reading, the designer of an online map can also set actions when the user clicks on a button or scrolls down the page. For example, the view can be progressively centered on a specific common point, a district, or simply switch from a 2D mode to a 3D view (flyto). Lastly, it is also possible to offer users views with several maps, either by synchronizing (synchronized map) to compare, or overlapping by swiping (swipe map), to visualize evolutions.

With regard to the forms of interaction with the data and in particular their properties, the main functions revolve around the management of layers via selection menus, sorting or filtering on the attributes of the data via buttons or bars, the search bar and the time bars (timeslider). Another form of interactivity is the display of information about geographic objects through popup windows, which can appear by clicking on a spatial object or by hovering over it. The visual query (brushing) is becoming more and more important, especially within dashboards, where it is possible to select map elements via a graph and vice versa. In this logic, it is also possible to display spatial objects by clicking or hovering over other objects in the vicinity or with similar attributes. There are also functions closer to those of GIS, such as importing data sets or data flows, the possibility of drawing on the map, measuring lengths or areas, or performing analyses (calculation of routes or isochrones, for example).

At the level of graphical rendering of data, the forms of interactivity essentially refer to logics where the display modalities are configured upstream by the designer of the online map, so as to optimize the reading and use of the online maps. It is thus possible to configure the appearance or disappearance of a layer, as well as its formatting, according to the zoom level. It is also possible to configure multi-scale data displays in order to map statistical indicators at different levels of aggregation (states, counties, tracts blocks), depending on the zoom level. Lastly, it is possible to arrange the layout of the data according to the zoom level of the map. The size of a point, the thickness of a line, the color of a polygon or the opacity of a spatial object can thus be adapted to the scale of visualization, so as to allow the user to read online maps more clearly, particularly avoiding the effects of superposition at smaller scales.

6.4. Overview of the Geoweb cartographic language

Beyond the characteristics of web maps in terms of their dynamic and interactive aspects, the challenges of the representation modes of geographic data raises a series of questions from the perspective of understanding cartographic communication on the Geoweb. The modalities, as well as the logic of formatting within web interfaces, refer to a cartographic language mobilizing elements and principles that are both traditional and most innovative, which should be examined.

6.4.1. *Back to the Geoweb semiology*

Cartography on the Geoweb takes form in an environment where the traditional rules of cartographic expression are no longer as prevalent as they once were in publishing cartography or GIS. Many web maps are based on computer graphics and web design, often to the detriment of basic cartographic principles and conventions. Examples of bad maps that have been produced and widely distributed on the web are common, both on generalist, institutional and press sites. Most errors are concentrated around information overload, overlay effects, lack of a legend or scale, and the use of unsuitable visual variables (Bielecka and Dukaczewski 2009). Beyond the benefit of showing some interesting cases to students, however, these errors are indicative of a new way of mapping the world that disregards established and taught rules of graphic representation.

This situation is not new; in the 1990s, Poidevin (1999) pointed out that with the sophistication of CAD and GIS tools, a good part of the maps produced came from computer graphics designers. Today, new web cartographers include other profiles such as web developers, data scientists or data journalists, for whom the map is one of several means of representing an ever-increasing amount of data (Usher 2019). Many of these audiences are unaware of the basic principles of graphics, how cartographic communication works, and the principles of graphic semiology (Griffin and Fabrikant 2012; Ipatow and Harvey 2019). Moreover, in their approach, they do not necessarily seek to produce rigorous maps. Instead, they explore data sets through maps in order to derive information and meaning from them in a comprehensive and exploratory approach.

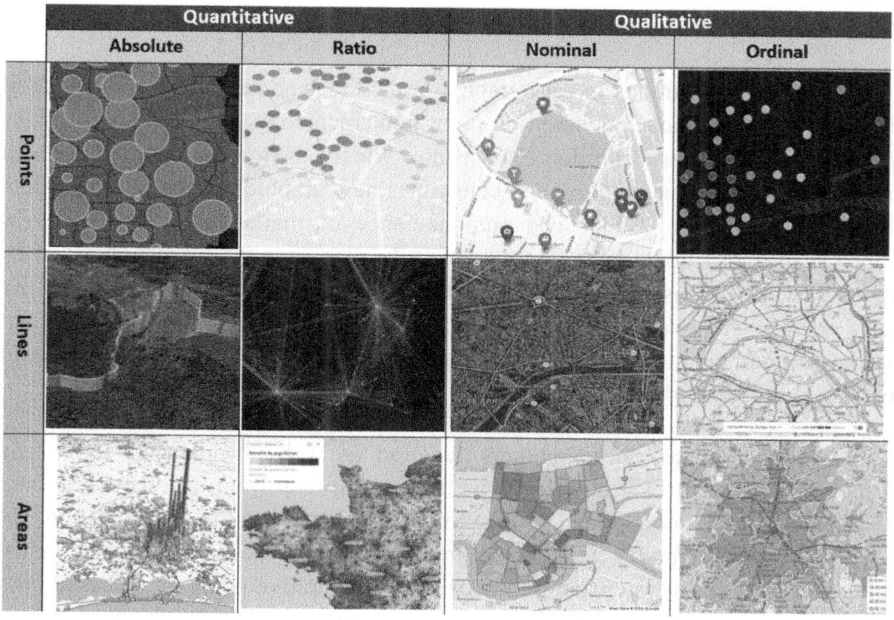

Figure 6.3. *Summary of Geoweb spatial data representation modes (according to Mericskay 2016). For a color version of this figure, see www.iste.co.uk/mericskay/cartographic.zip*

On the Geoweb, it is indeed more about visualizing data on a map than revealing differences or making cross-references. In fact, there is a predominance of visual variables such as color and shape (Mericskay 2016; Ipatow and Harvey 2019). Despite the development of a cartographic culture based on fairly simple and standardized modes of representation, however, there is a rich palette of representation modes for transcribing difference, order, or proportionality relations between qualitative and quantitative data on web maps (Figure 6.3). The cartographic language of the Geoweb is based on a system of signs that heavily relies on traditional conventions and trends, similar to Bertin's graphic semiology (Béguin and Pumain 2017). Moreover, in the discourse, it is no longer a question of graphic semiology, but rather of data-driven cartography, as put forward by the major publishers of web mapping solutions. The development of mapping on the Geoweb is based on a reinterpretation of cartographic rules and principles that were established more than 50 years ago, but reinvested in a different sociotechnical context where practices, users and tools have multiplied and diversified.

6.4.2. *The hegemony of map markers*

One of the main trends in cartography on the Geoweb lies in the prevalence of figurative cartographic signs based on point implementation. This hegemony of the map pin (Singleton and Brunsdon 2014) is now widely anchored in practices, as well as in the collective imagination, and constitutes new graphic vocabulary in modern cartography (Wallace 2011). The preponderance of markers refers to the very principle of transactional cartography, where the map answers a question asked by the user (where is this address? What is the nearest restaurant?). At the core of our daily practices, the web map is positioned as a spatial extension of the search bar, and has the particularity of presenting the results of a query on a map, in the form of markers (Loidl 2011).

Within web maps, these markers function as web objects (interactive and clickable) and refer to various web content, itself coming from different sources in an aggregation logic (Boulos 2008). In fact, the modes of representation are essentially based on points with simple graphics, without the use of complex visual variables, allowing hierarchization, for example. Figurative point signs, based on a visual shape variable, are generally very easy to understand by inexperienced map readers, even without the use of a legend (Boulos 2008). The almost systematic use of pictographs, as a schematic graphic representation with a sign function, provides readers with familiar markers with shared meanings (Ipatow and Harvey 2019).

It is interesting to observe that the ways in which cartographic markers are represented are changing over time. From colored pins to dynamic and interactive markers with labels, to pictograms, the semiology of point features is becoming more complex (Figure 6.4). Geoweb cartographers have a number of sign libraries at their disposal to format the multitude of point data, such as Map-icon, Mappity or Maki-icons. Available in vector formats, these markers are based on variations in shape (pictograms) and color, and even size.

One of the current trends is the enrichment of markers with labels, called "badges". Uber, which is very active in the development of new online mapping tools, is proposing a marker syntax based on the representation of multiple pieces of information within a single map object. This system, which combines markers (based on size, color and shape variables) and labels, allows on-the-fly display of various information in point form (place names, prices, travel time, vehicle identifiers and notes) within online maps of mobile applications.

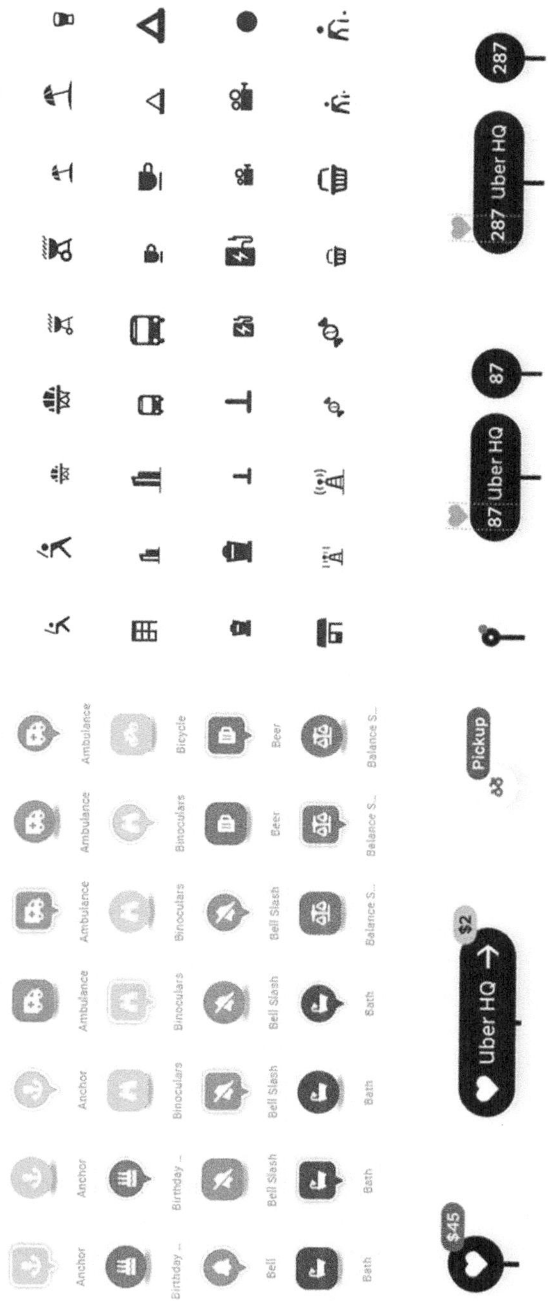

Figure 6.4. *Examples of map markers (sources: Mappity, Maki-icons, Uber). For a color version of this figure, see www.iste.co.uk/mericskay/cartographic.zip*

6.4.3. *Thematic cartography representation modes*

In addition to the representation of markers on base maps, the Geoweb is also characterized by the importance of thematic cartography. Many web maps present all kinds of localizable phenomena, be it qualitative or quantitative (demography, election results, real estate prices, Covid-19 pandemic, etc.).

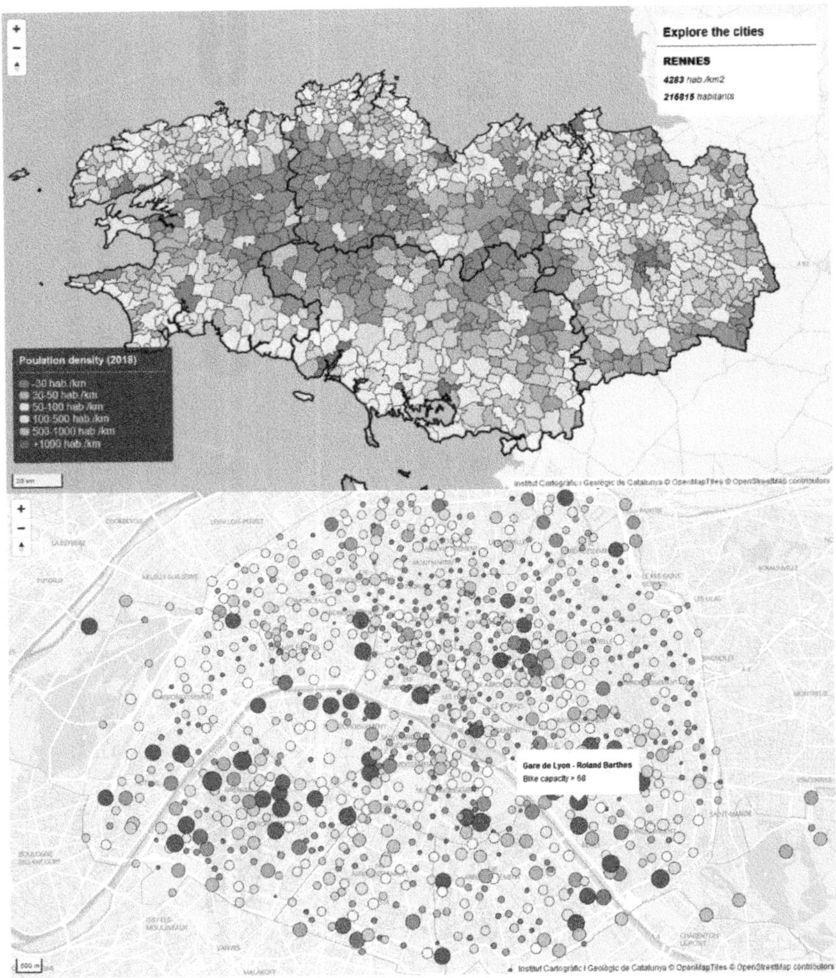

Figure 6.5. *Examples of a choropleth map (leaflet) and a proportional symbols map (KeplerGL) (author: Mericskay) (sources: INSEE, OSM). For a color version of this figure, see www.iste.co.uk/mericskay/cartographic.zip*

First of all, the choropleth map (Figure 6.5), which has a zonal layout, is prominent. Although a very popular form of cartographic representation, maps in solid colors are based on a variety of color palettes and often lead to semiological errors (poor discretization, representation of stock variables, large number of classes, etc.). However, the particularity of web choropleth maps lies in their dynamic and interactive aspect. On the one hand, they are not fixed, but are quickly generated by users, who can easily modify the scale of visualization, the data to be displayed and their formatting modalities (MacEachren et al. 2008). On the other hand, the display of the properties of the represented variables in information windows allows the user to examine the spatial entities of the map and to compare them in a very simple way. In addition, and in the logic of the dashboards that have previously been discussed, the presence of graphics offers users other angles of reading and exploring the data represented using an array of representation modes.

Another "traditional" cartographic mode that is present within the Geoweb is the proportional maps, based on variations in size that allow the representation of populations, measurements, counts, etc. (Figure 6.5). Proportional symbols (mostly circles) are often achieved through blending effects from a map design perspective. However, there are several important limitations to their proper understanding. First, few proportional symbols maps offer legends, as the instrumentation does not handle this aspect very well. In most cases, the design of a legend has to be done in an artisanal way (image or HTML element). Another limitation is that the methods for calculating the sizes of the tools are essentially based on graduated (class) and non-proportional symbols logic, which can cause representation problems. Lastly, the size of the symbols is often fixed, and when the user switches to a larger scale, the size of the symbols does not change, thus leading to an overlay of geographical objects, which become more numerous and limit the reading of the maps.

6.4.4. *Spatial representation of big data*

Beyond reinvesting in already well-known cartographic representation modes, one of the main challenges for cartography on the Geoweb is to propose both methodological and technical solutions in order to represent and display thousands, even millions of spatial entities in a fluid, readable and understandable way (Gaffuri 2012; Robinson et al. 2019). This issue of volume mainly concerns point data, which are increasingly numerous and varied. Conventional methods of thematic cartography that mobilize visual variables of size, color or shape are unable to meet these new geovisualization needs, due to the sheer volume of data (Netek et al. 2019). Displaying thousands of points as markers introduces the problem of overlapping data, which limits their identification and generally decreases cognitive interpretation capabilities (Ježek et al. 2017). In addition, there is the issue of graphical rendering performance, which requires rethinking the representation modes to minimize the size of the transmitted data and thus make the display smoother (Fürhoff 2019; Tang 2020). In response to these needs, several approaches, stemming from information visualization techniques and methods, are being increasingly mobilized, making it possible to interpolate, aggregate or summarize large volumes of data, and ultimately facilitate their interpretation.

6.4.4.1. *Dot maps*

The dot map technique, used for a long time in thematic cartography, is becoming increasingly popular on the Geoweb. A dot map is a type of thematic map that uses a point sign, often of uniform size, to visualize the geographic distribution of a phenomenon. In response to the proliferation of spatial data, especially point data, and the possibilities for a dynamic and interactive display of large volumes of geographic objects, point maps are proliferating around a variety of thematic objects such as immigration, educational attainment, language and real estate prices (Walker 2018). However, the most dynamic theme is racial dot maps, which aim to represent the reasoning behind ethnic spatial distributions, particularly in North American cities (Seguin et al. 2017).

Point maps are based on a visual distribution that better highlights the spatial distribution of a phenomenon. They offer another view of the territories by spatializing large volumes of point data, allowing the highlighting of geographical rationales (concentration, regular dispositions). In this type of geovisualization, the privileged visual variable is the color that adheres to a logic of categorization or graduation of the variables to be represented.

We can distinguish three main types of point maps, referring to different logics of spatial modeling of the represented information. In the first type, each point can represent an object or event that is individually modeled by its location (one Wikipedia page, one tweet, one building, one crime, one tree, etc.), as shown in Figure 6.6. In the second type, each point on the map does not represent an individual data, but refers to aggregated data such as a population, or a number of voters. The data are based on predefined geographic divisions such as census tracts, in which the data on individuals have been aggregated as statistical variables. The method then consists of randomly placing the number of points related to the variable in each area. The third type is dot maps that represent aggregated data, but where each point on the map refers to the centroid of polygon. In this case, the expected effect of point density will only be effective if the number of points, and thus of input polygons, is very large, such as at regional or national scales to represent sociodemographic data.

Figure 6.6. *Dot map of houses and buildings in the inner city of Île-de-France (author: Mericskay; sources: ARCEP, OSM). For a color version of this figure, see www.iste.co.uk/mericskay/cartographic.zip*

6.4.4.2. *Spatial grouping (clustering)*

The spatial grouping of markers into clusters is very popular with computer scientists and web developers alike and plays an important role in Geoweb cartography (Fürhoff 2019). This method, based on the analysis of distances and densities of a set of points, allows a map that is saturated with points to be made more readable, by grouping them into homogeneous clusters and displaying their number (Meier 2016). With increasingly powerful hardware environments and browsers, it is possible to process millions of spatial markers in real time and offer very fast display load times on the client side (Netek et al. 2019).

In terms of graphical rendering, there is a trend toward more complex forms of cluster representation (Figure 6.7). In order to increase efficiency, adding visual variables (size, shape, color) to clusters helps users to identify, classify and compare entities on the map. The most common form consists of introducing a new visual entity into the map that differentiates clusters (usually circles) from single markers, and is labeled with the number of aggregated markers. Other techniques allow for a combination of size and color variation in the representation of clusters, so as to better differentiate the volume of aggregated markers. It is also possible to integrate quantitative variables associated with the markers, which can themselves be taken into account in the aggregation (sum, average) and displayed within the labels. Lastly, more complex methods are starting to emerge, such as those that make it possible to quickly produce clusters based on proportional diagrams. In this case, only the aggregation radius is configurable, the underlying clustering methods having already been pre-configured within the JavaScript library.

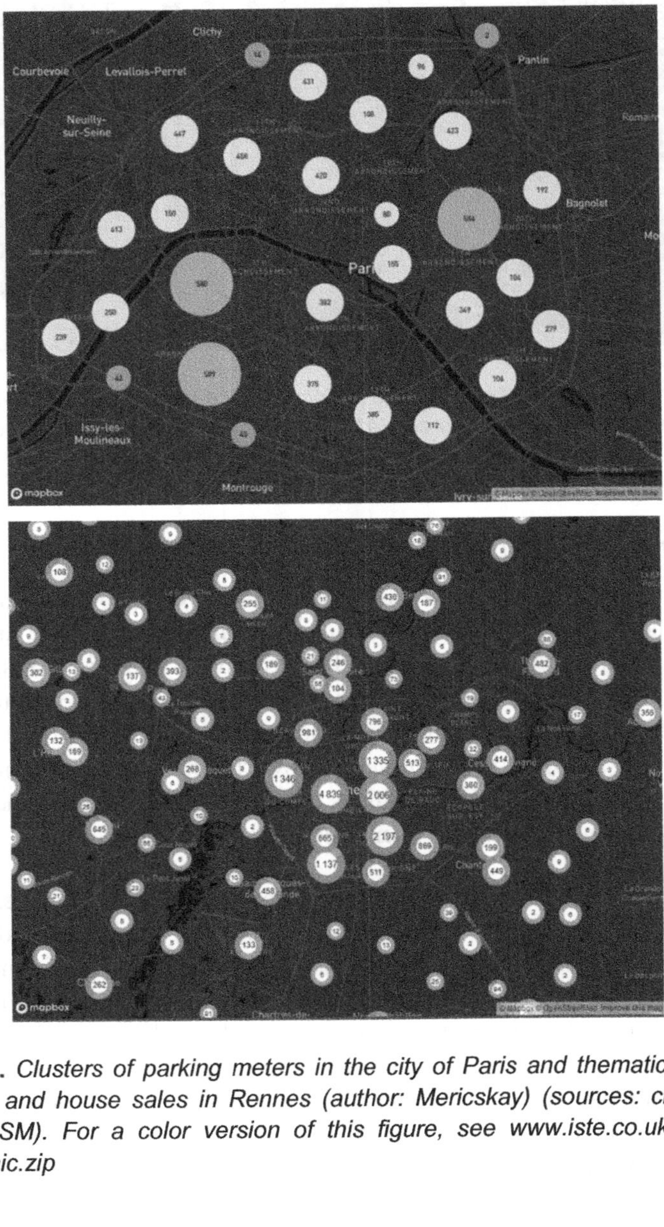

Figure 6.7. *Clusters of parking meters in the city of Paris and thematic clusters of apartment and house sales in Rennes (author: Mericskay) (sources: city of Paris, DGFIP, OSM). For a color version of this figure, see www.iste.co.uk/mericskay/cartographic.zip*

6.4.4.3. *Grids*

Another technique that is increasingly used on the Geoweb is spatial aggregation of grid-based data, also known as binning. This method, which is well known in spatial analysis, proposes to summarize and represent large spatial data within regular and homogeneous spatial units (Amini et al. 2011; Kiefer 2015). This particular way of partitioning space, whose basic unit is the mesh, can take different forms in terms of the geometries used (tiles, hexagonal meshes, diamonds or rectangles). By freeing itself from administrative divisions, this approach, which is based on a homogeneous tessellation of space, makes it possible to highlight the spatial distribution trends of a phenomenon.

Visually, the representation in the form of a tessellated grid of regular surface entities offers an easier reading of the information (Kiefer 2015). Furthermore, the added value of this method lies in the ability to perform a set of functions to statistically summarize the semantic features of the input data, as illustrated in Figure 6.8 (count, average, sum, etc.). In addition, one of the most significant innovations is the dynamic aspect of the grids and, by implication, the aggregations performed. Generated on-the-fly, the granularity, the shape of the grids, as well as the statistical summary operations can be adapted to the scale of visualization and thus propose grids according to grainy or very fine resolutions, depending on the data or the targeted objectives. This method is particularly relevant for the exploration of spatial data, characterized by numerical attributes. However, the setting of the size as well as the shape of the grids is not always customizable and can even be impossible in some cases, thus limiting the interpretation of these spatial representations.

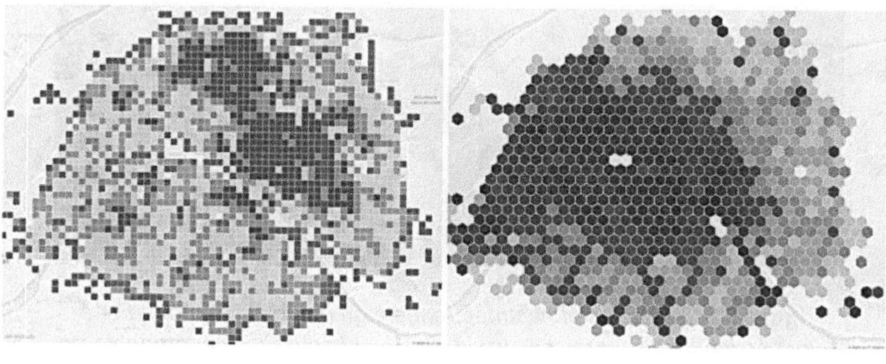

Figure 6.8. *Counting the number of Airbnbs (grid) and the average price of Airbnbs (hexagonal grid) in Paris (author: Mericskay) (sources: Inside Airbnb, OSM). For a color version of this figure, see www.iste.co.uk/mericskay/cartographic.zip*

6.4.4.4. *Heatmaps*

Unlike cluster and grid approaches, which produce a discrete representation of the data, heatmaps[7] offer a continuous representation of a point data set through the estimation of kernel densities (Ježek et al. 2017). Heatmaps provide an intuitive visual overview of the intensity of a phenomenon, with respect to a spatial region (Figure 6.9). It is thus possible to quickly identify areas of high density (hotspots), distribution, as well as the spatial organization of concentrations in a data set (DeBoer 2015; Netek et al. 2019). On the Geoweb, heatmaps are dynamic and adapt to the scale of visualization. In terms of formatting, the tone range, color chart, intensity and opacity of heatmaps are mostly customizable. However, generally speaking, this raster-based representation mode does not offer much variety in terms of formatting, in that only the visual color variable is customizable (Meier et al. 2016).

This mode of representation, which is very popular on the Geoweb, has other limitations, particularly in terms of configuring the visual rendering of the maps. Unlike GIS tools, where the user can choose different methods and configure them as desired, in web mapping tools, few parameters can be configured to produce clear and relevant heatmaps (radius, interpolation method, weighting variable). Although heatmaps are visually effective, the proposed representations can be incorrect or even misleading. This technique should be used with caution, because there is often little guidance on how these maps are constructed and their designers often have little control over their configuration (DeBoer 2015).

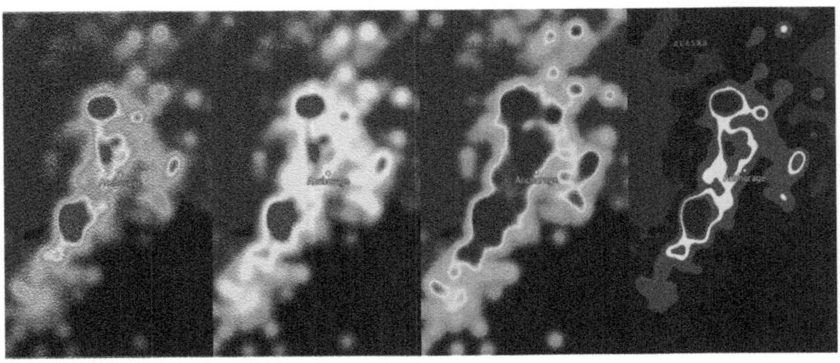

Figure 6.9. *Examples of heatmap renderings (source: Mapbox). For a color version of this figure, see www.iste.co.uk/mericskay/cartographic.zip*

7 A heatmap is a raster grid, for which each pixel is assigned a measurement value corresponding to the density of points in and around each pixel.

The use of dot maps, heatmaps, grids and clusters is an important part of the cartographic language of the Geoweb. Beyond the technical opportunities to represent large volumes of spatial data, the question of visual efficiency in terms of understanding the cartographic message arises. However, we should not lose sight of the fact that the objective of these types of maps is not to provide fixed representations for analysis or communication purposes, but rather to provide interfaces to explore masses of spatial data in a dynamic and interactive way. Moreover, these approaches based on aggregation do not aim to represent geographical data according to an individual approach, but rather to show spatial structures and concentration logics, in a multi-scale perspective.

6.4.5. *The representation of spatial data in three dimensions*

The development of WebGL as a standard for 3D visualization on the web is an important technical innovation in the development of web maps (Krämer and Gutbell 2015; Farkas 2017; Mericskay 2021). This specification for web browsers allows complex 3D graphical elements to be dynamically displayed, created and managed within a web browser, without the use of additional modules (Farkas 2019). WebGL transforms web maps, especially by getting rid of the classic north-facing planisphere view. In addition to the classic pan and zoom functions, it is also possible to rotate and tilt the maps. This evolution of interactive viewing brings about new ways of thinking about web maps and ways of representing spatial data through the use of 3D extrusion. This shift to the third dimension induces new forms of data perception, which call for different interpretative capabilities than those of two-dimensional maps (Ware and Plumlee 2005).

The building repositories, which are more exhaustive, as well as being more geometrically and semantically precise (number of floors, height, etc.), are increasingly visualized in three dimensions on the Geoweb. With WebGL, it is easy to extrude the height of buildings to show the volume of cities in a more realistic way. The idea is to provide 3D base maps, where the user can move around and overlay other data. In terms of formatting, more and more geovisualizations propose applying categorizations or color scales to buildings, so as to represent different information such as height, number of buildings, age, or current zoning (Figure 6.10).

Figure 6.10. *3D extrusion of building heights in Paris (author: Mericskay; sources: city of Paris, OSM). For a color version of this figure, see www.iste.co.uk/mericskay/cartographic.zip*

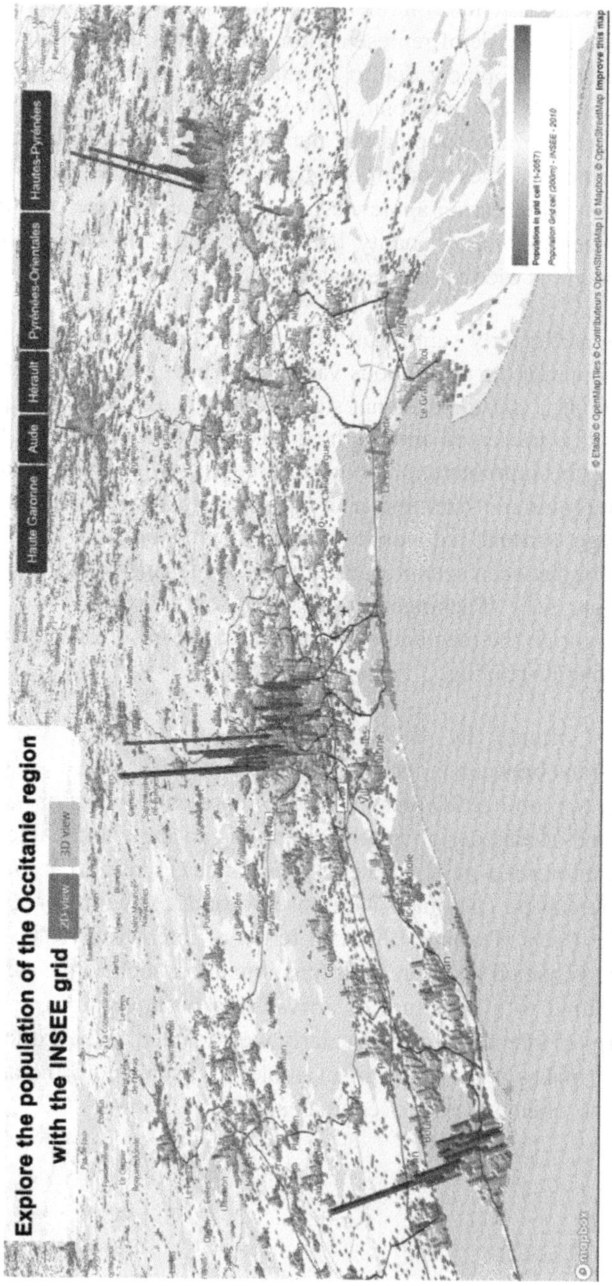

Figure 6.11. *3D extrusion of the population of the 200 m grid data (author: Mericskay; sources: INSEE, OSM). For a color version of this figure, see www.iste.co.uk/mericskay/cartographic.zip*

3D thematic maps, which allow three-dimensional representation of statistical data, are also experiencing a renewal in their uses. Having been explored for a long time by cartographers, this mode of representation, which was traditionally difficult to produce, is today in development due to technical advances in 3D display. The principle is simple: extrude 2D polygonal entities (administrative contours, grids) according to a statistical variable (population, number of accidents, electoral results, etc.). Each polygon thus becomes a prism of height, proportional to the value of an attribute (Rase 2009). This cartographic design technique has the advantage of representing differences in proportionality between spatial units, according to an approach in which height is used to differentiate them (Jégou 2007).

This approach is particularly significant in thematic mapping because it allows an alternative representation for continuous variables, which are most often shaped by a variation in size of a two-dimensional sign. In the 3D extrusion of a polygon, the shape of each unit is preserved and the height is used as a visual variable. It is therefore possible, at the representation level, to combine two visual variables for the same spatial unit (color and size): either to represent a single variable in order to reinforce the message (virtual redundancy through extrusion and color) and to promote differentiation between spatial entities of the same color class, as shown in the example in Figure 6.11 (Pachinger et al. 2016), or to express two different variables in order to enrich the map and cross-reference the information, albeit with the risk of complicating its reading.

The possibilities offered by WebGL also renew the ways of spatially representing interactions between an origin-destination matrix[8]. Based on a 2D link map, it is now easier to mobilize 3D arcs in order to represent connections and flows on a map. This method of map design involves applying an angle to the line, so as to turn it into a curve, giving it a more aesthetic appearance (Figure 6.12). In 3D, this curvature of the line can be combined with an elevation, allowing for a different representation of the flows (Dorling 1991). In terms of interactivity, this mode of representation is accompanied by interesting functionalities, such as the selection of a subset of data relating to a departure or arrival area by hovering over the map (brushing). This form of interaction is particularly well adapted to the exploration of flow data because it makes it possible, for example, to visualize all the associated origins or destinations when hovering over an area.

8 For a detailed overview of flow mapping (principles, forms and challenges), see Chapter 4.

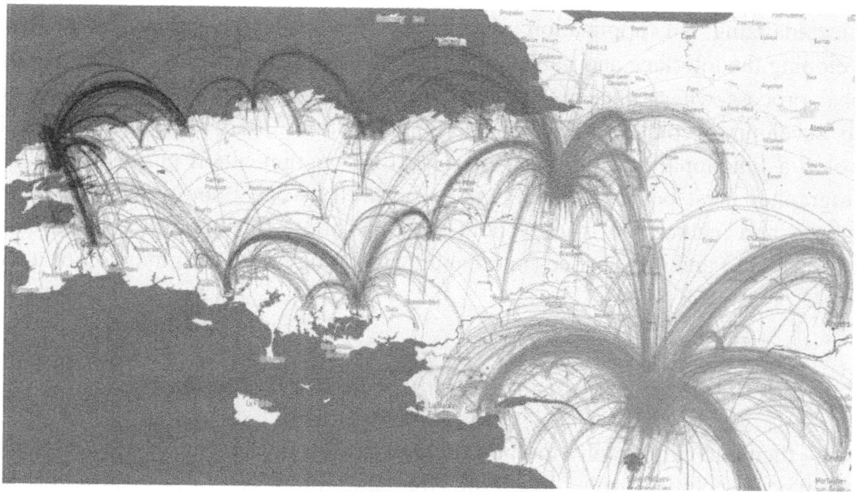

Figure 6.12. *Extruded arcs representing commuter carpooling trips (author: Mericskay; sources: carpooling evidence log, OSM). For a color version of this figure, see www.iste.co.uk/mericskay/cartographic.zip*

With WebGL, it is now easier to produce online 3D maps with efficient graphics rendering and loading times, offering users original, more interactive and very fluid representations. However, this approach has limitations, especially in terms of readability and interpretation of the entirety of the data represented. For example, some extruded data may obstruct parts of the map, thus limiting a complete reading of the data on the represented territory. Depending on the number of spatial units represented, the map can also be more or less expressive. An excessive number of extruded units lead to a very fragmented image that is sometimes difficult to interpret (Cauvin et al. 2010). Moreover, the ability to orientate within the maps can be affected by a lack of ambiguous and conventional landmarks. Interactivity offers an answer to this problem by allowing users to move, rotate, tilt the view and change scale in order to be able to explore and navigate within these representations.

6.5. The importance of technology for cartographic communication on the Geoweb

It is customary to consider technical instrumentation as secondary in mapping processes, insofar that the method matters more than the tool (Lambert and Zanin 2016; Béguin and Pumain 2017). While this stance appears legitimate in the context of static map design, with each cartographer relying on his or her own toolbox (GIS, CAD, DAO), in the context of web mapping, it is quite a different logic.

Conceptualizing an application, formatting the data, managing interactivity, developing the interface and publishing a map online are operations that are part of a global process where data preparation, spatial analysis, cartography, hosting and web development practices are all intertwined (Smith 2016; Tolochko 2017). To develop a functional and efficient interface for communication purposes, the web cartographer must possess multiple skills, as well as call upon a set of technical solutions and methods to manage both the back office (hosting, API, cloud-based platforms, etc.) and front office (interfacing, symbology, interactivity, etc.) of a web map.

6.5.1. *The thematic cartography toolbox on the Geoweb*

The thematic cartography toolbox on the Geoweb is characterized by numerous and diverse technical solutions, both free and proprietary, and more or less accessible. Several comparative analyses help to understand this technical ecosystem, as well as the advantages and disadvantages of the plethora of online solutions, especially for thematic mapping (Roth et al. 2015; Farkas 2017). Faced with this diversity of tools, it is possible to divide them into two main categories: (1) solutions based on graphical interfaces, which allow all types of audiences to produce online maps in a rather simple way, without lines of code and (2) programming libraries that offer the possibility of creating web maps with more flexibility, but require specific skills in web development.

Web mapping environments that are based on graphical interfaces have undergone significant development over the last 10 years. These applications, based on the principle of SaaS (Software-as-a-Service), allow users to import and spatialize data, load base maps, mapping data and distribute maps without any line of code, and without worrying about hosting issues (Panek and Burian 2020). On the one hand, there are online versions of GIS solutions such as ArcGIS online (ESRI), GEO (Business Geographic) or CARTO, which, in addition to cartography, offer spatial analysis functionalities (Fast and Hossain 2020). These intuitive applications, which come from the world of GIS and are quick to learn, are not free, because they are hosted on private servers. Another recent interface application (Figure 6.13), known as KeplerGL (She 2020), is based on different logic than the solutions mentioned above (open source, no data hosting, visualization-oriented, etc.). Developed by the company Uber, and aimed more towards data scientists, this web application makes it possible to explore and map large and complex geographical datasets by mobilizing a wide range of visualization methods of spatial big data (dot maps, 3D extrusion, heatmaps, clustering, grids, spatiotemporal animation, etc.).

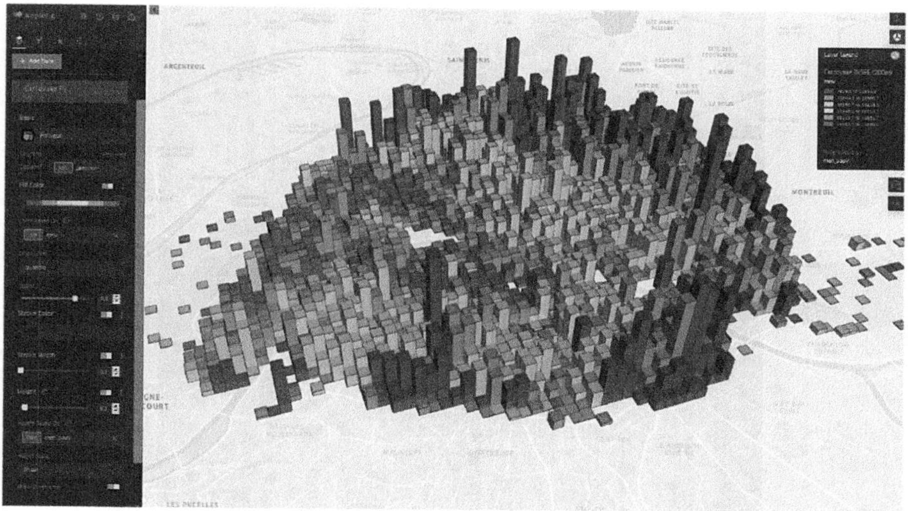

Figure 6.13. *KeplerGL web application interface (source: Uber). For a color version of this figure, see www.iste.co.uk/mericskay/cartographic.zip*

For their part, JavaScript programming libraries are becoming more and more important in web mapping. They are evolving very quickly with technological advances such as web map services, vector tiles and WebGL. Unlike interfaced applications, these libraries (on the client side) that allow maps to be designed within web pages require development work, not only in JavaScript but also in HTML and CSS (Figure 6.14). After a decade of strong momentum where solutions (proprietary and open source) have multiplied, there has recently been a refocusing on three main libraries: MapboxGL, Leaflet and OpenLayers, which concentrate the majority of current usage (Netek et al. 2019; Zunino et al. 2020).

On the one hand, the OpenLayers and Leaflet libraries are part of very dynamic open source projects and communities, allowing them to evolve rapidly and to develop new functionalities. However, they have some limitations such as the fact that they are not very well adapted to cartography, they do not yet manage 3D displays or are not very well optimized for the display and representation of large data. On the other hand, Mapbox, which is under commercial license, is more focused on design, massive data representation and mobile cartography. Recently, a new generation of online mapping libraries that is based on WebGL technologies has started to emerge. Developed by large geospatial companies, these libraries, like ArcGIS API for JavaScript (ESRI), Harp.gl (HERE) or deck.gl (UBER), specialize in the geovisualization of complex spatial data sets (large, 3D or spatiotemporal).

```
    // DOTMAP GRADUEE TAILLE + COULEUR

map.addLayer({
        'id': 'combine',
        'type': 'circle',
        'source': 'DVF',
        'source-layer': 'DVFRM-3fc18X',
        'layout': {'visibility': 'visible'},
        'paint': {'circle-color': {property: 'Prixm2OK',type: 'exponential',
                stops: [
                    [0, '#1a9850'],
                   [2000, '#66bd63'],
                   [2250, '#a6d96a'],
                   [2500, '#d9ef8b'],
                   [2750, '#fee08b'],
                   [3000, '#fdae61'],
                   [3250, '#f46d43'],
                   [3500, '#d73027'],]},
                'circle-radius': { property: 'PrixOK', stops: [
                [{zoom: 11, value: 10000}, 0.3],
                [{zoom: 11, value: 1600000}, 3.5],
                [{zoom: 18, value: 10000}, 3],
                [{zoom: 18, value: 1600000}, 15] ]},
                'circle-opacity': 0.8}

    });
```

Figure 6.14. *Examples of JavaScript syntax (MapboxGL) to apply color and size variation to point entities. For a color version of this figure, see www.iste.co.uk/mericskay/cartographic.zip*

Some JavaScript mapping libraries are also available in the main (spatial) data analysis languages such as R or Python. Available through libraries, it is possible to design web maps directly from these working environments, without going through web development. For the R language[9], for example, the Leaflet for R or Shiny packages allow the production of web maps and dashboards directly from R Studio. As far as the Python language is concerned, libraries such as folium or pydeck offer features to produce 2D or 3D web maps, directly within the processing chains. This new way of producing web maps from development environments dedicated to data analysis is not just anecdotal. It opens up many new perspectives in cartography and web mapping to new audiences who are fond of interactive visualizations and, more globally, raises questions about the role of (web) maps in the development of spatial data science.

6.5.2. *Increasing technicalization of practices*

In a manner quite similar, if not more present than with GIS software, on the Geoweb, mappers are both guided by technical solutions and forced to use the sign systems provided to compare, differentiate, order and generally visualize and transmit/communicate information. Web mapping is both a semiotic and a procedural tool that offers a vocabulary of signs and functions (especially interactivity) that is part of both graphic design and programming (Rieder 2009). In theory, the possibilities of mapping spatial data on the Geoweb appear to be unlimited, as the tools, technologies and data are so numerous. However, in practice, if the user wishes to explore forms of representation other than those proposed, they must have knowledge and skills in programming that go beyond the functionalities that are natively present within the solutions (Roth et al. 2014).

The main problem in these increasingly technical environments is designers have little room to maneuver in the configuration of web maps (discretization methods, choice of colors, settings of aggregations or interpolations, interactive features, legend design, etc.). Furthermore, the visual rendering of representation modes such as clusters or heatmaps often lacks meaning, as the proposed aggregations and interpolations are the result of automated processing, where cartographers have little control over the settings of aggregations and formatting (Furhoff 2019).

Despite a great diversity of spatial data mapping possibilities, we are witnessing a certain standardization of representation modes, like cluster maps or heatmaps, that are increasingly present on the web. As Zanin and Ysebaert (2017) remind us, innovation in cartography can only really emerge if the means given to the cartographer allow him to let his creativity express itself. However, on the Geoweb, the tendency is to implement generic tools that have already been preconfigured (interfaced solutions, JavaScript libraries, APIs), where map designers are, for the

9 Chapter 3 presents the technical ecosystem of R for thematic cartography in more detail.

most part, constrained by technical frameworks and increasingly standardized modes of representation. Greater freedom of design for web maps necessarily requires web programming skills (HTML, CSS and JavaScript). The result is a never-easy balance between the creative aspect and the configuration aspect of web maps (Bostock and Davies 2013). Producing web maps is not just a challenge for cartographers, as it also depends on cooperation and communication between programmers and mappers (Moser and Koslitz 2016).

A major challenge in the technicalization of web mapping lies in teaching future and current professionals in these tools and logics (White and Powell 2019). Issues of (geo)digital literacy and learning to code emerge as keys to better appropriation and more effective use of instrumentation in cartographic communication. This challenge is exacerbated by the fact that the majority of web mapping libraries are developed outside of the discipline or refer to (design) goals that do not necessarily conform to cartographic principles (Poorthuis et al. 2020). Many applications are indeed designed by (non-mapping) computer developers, who propose more and more modes of representation without necessarily mastering the codes, nor questioning the possible forms of use (Eckert and Jégou 2008). As Cauvin (2016) points out, the representation modes of spatial data chosen by Internet users, whether or not they are trained in cartographic principles, are those of designers who have integrated them into their solutions. It is therefore up to designers to be competent, but also responsible and proactive in proposing symbologies that are adapted to the properties of contemporary spatial data, to the new profiles of cartographers and to the contemporary uses of web maps.

6.6. Conclusion and perspectives

This chapter clearly shows that, on the Geoweb, the representation modes of geographic information (already disrupted by GIS), as well as the construction of a (web) map, are changing profoundly. The new uses of web maps are accompanied by new graphic conventions that are specific to distributed computing environments and new users, but above all adapted to the growing volumes of data.

On the Geoweb, the map is first and foremost a tool for searching and visualizing spatial data, not a static medium for transmitting a message. This ease to quickly represent complex and large spatial data in diversified forms, as well as to offer users dynamic and interactive web environments, is accompanied by a "black box" effect. As cartography becomes more and more popular with a wider audience, it loses its rigor, in that the cartographer no longer necessarily masters all the steps in the construction of his map and, by extension, the message he wishes to convey. The esthetic aspect and the design are becoming more and more significant, to the detriment of the elementary rules of thematic mapping.

However, despite a dominant discourse focused on renewal and innovation around methods and tools, the development of "data-driven" web mapping does not really mark a breakthrough in the way data are represented through maps. The new generation of tools borrows heavily from traditional cartographic languages and sign systems, as well as from well-known methods from spatial analysis and information visualization. One of the main innovations in web mapping today is the progress in graphic processing and rendering capabilities, such as vector tiling and WebGL. The combination of these two techniques allows for the optimization of the transmission and display of large volumes of spatial data, while simultaneously facilitating the use of complex representation modes such as 3D or aggregation methods. Some operations that can take several minutes or even hours in GIS are now completed in a few seconds on the Geoweb.

Advances in interactivity between the user, the data and the representations constitute another major innovation, particularly in terms of cartographic communication. The forms of interaction on the Geoweb are increasingly varied, both in terms of data (selection, filtering), views (pan, changing scale, tilting, rotating) and representation modes (choice of base map, customization of colors and change of aggregation modes). This evolution positions the web map as an instrument for navigation and exploration of masses of spatial data, where readers have the ability to produce their own maps.

As a result of its intrinsic characteristics, cartography has always been very sensitive to technological progress. With the Geoweb, these transformations are profound and lasting. It seems very important that the map, as a means of communication, retains its specificities and a certain rigor, so that it does not become a simple mode of visualization of data among others. In order to meet this challenge, professional mappers must invest more in this mutation, from a methodological, technical and thematic point of view. By relying on the current logic and tools for development and collaborative documentation (GitHub, GitLab, Observable, etc.), web cartographers can make the solutions evolve by proposing new functionalities, or even by developing and sharing new tools. This is the aim of many open-source web mapping projects: to encourage creation and innovation around the representation of geographic data and, by extension, cartographic communication.

6.7. References

Amini, A., Wah, T.Y., Saybani, M.R., Yazdi, S. (2011). A study of density-grid based clustering algorithms on data streams. *8th Conference on Fuzzy Systems and Knowledge Discovery*, 3, 1652–1656.

Antoniou, V., Morley, J., Haklay, M.M. (2009). Tiled vectors: A method for vector transmission over the web. In *Web and Wireless Geographical Information Systems*, Carswell, J.D., Fotheringham, A.S., McArdle, G. (eds). Springer, Berlin.

Battersby, S.E., Finn, M.P., Usery, E.L., Yamamoto, K.H. (2014). Implications of web Mercator and its use in online mapping. *Cartographica*, 49(2), 85–101.

Batty, M., Hudson-Smith, A., Milton, R., Crooks, A. (2010). Map mashups, Web 2.0 and the GIS revolution. *Annals of GIS*, 16(1), 1–13.

Béguin, M. and Pumain, D. (2017). *La représentation des données géographiques : statistiques et cartographie*. Armand Colin, Paris.

Beukelaar, I.T.Y. (2018). Cartographic implications of Vector Tile technology. Geomatics Master's Thesis, University of Utrecht, Utrecht.

Bielecka, E. and Dukaczewski, D. (2009). Web mapping errors and their typology. In *24th International Cartographic Conference*. Santiago.

Bostock, M. and Davies, J. (2013). Code as cartography. *The Cartographic Journal*, 50(2), 129–135.

Boulos, K. (2008). Principles and techniques of interactive Web cartography and Internet GIS. In *Manual of Geographic Information Systems*, Madden, M. (ed.). American Society for Photogrammetry and Remote Sensing, Bethesda.

Cauvin, C. (2016). La sémiotique en cartographie au XXIe siècle ? *Cartes et géomatique*, 229/230, 17–31.

Cauvin, C., Escobar, F., Serradj, A. (2010). *Thematic Cartography 2: Cartography and the Impact of the Quantitative Revolution*. ISTE Ltd, London, and John Wiley & Sons, New York.

DeBoer, M. (2015). Understanding the heat map. *Cartographic Perspectives*, 80, 39–43.

Dorling, D. (1991). The visualisation of spatial social structure. PhD Thesis, Newcastle University, Newcastle.

Dubrava, T. (2017). Design of a multi-scale base map for a tiled web map service. Master's Thesis, University of Twente, Enschede.

Eckert, D. and Jégou, L. (2008). Quel planisphère de référence pour Google Maps ? *Mappemonde*, 92 [Online]. Available at: http://mappemonde-archive.mgm.fr/num20/internet/int08401.html.

Elwood, S. (2011). Geographic information science: Visualization, visual methods, and the geoweb. *Progress in Human Geography*, 35(3), 401–408.

Farkas, G. (2017). Applicability of open-source web mapping libraries for building massive Web GIS clients. *Journal of Geographical Systems*, 19(3), 273–295.

Farkas, G. (2019). Hardware-accelerating 2D web maps: A case study. *Cartographica*, 54(4), 245–260.

Fast, V. and Hossain, F. (2020). An alternative to desktop GIS? Evaluating the cartographic and analytical capabilities of WebGIS platforms for teaching. *The Cartographic Journal*, 57(2), 175–186.

Few, S. (2013). *Information Dashboard Design: Displaying Data for At-a-Glance Monitoring.* Analytics Press, Berkeley.

Fürhoff, L. (2019). Rethinking the usage and experience of clustering markers in web mapping. *PeerJ Preprints*, 7, 278–583.

Gaffuri, J. (2012). Toward web mapping with vector data. In *Geographic Information Science*, Xiao, N., Kwan, M., Goodchild, M.F., Shekhar, S. (eds). Springer, Berlin.

Griffin, A.L. and Fabrikant, S.I. (2012). More maps, more users, more devices means more cartographic challenges. *Cartographic Journal*, 49(4), 298–301.

Griffin, A.L., Robinson, A.C., Roth, R.E. (2017). Envisioning the future of cartographic research. *International Journal of Cartography*, 3, 1–8.

Haklay, M., Singleton, A., Parker, C. (2008). Web mapping 2.0: The neogeography of the GeoWeb. *Geography Compass*, 2(6), 2011–2039.

He, S. (2020). Data-driven urbanism: The balance between spatial intelligence and design craftsmanship. *Architectural Design*, 90(3), 86–93.

Ipatow, N. and Harvey, F. (2019). Bertin's graphic variables and online map makers: An empirical study of maps produced by prosumers and cartographers. *Cartographica*, 54(4), 233–244.

Jégou, L. (2007). La troisième dimension en cartographie statistique, des cartes en prismes imprimées aux modèles 3D interactifs. *Mappemonde*, 86 [Online]. Available at: http://mappemonde-archive.mgm.fr/num14/articles/art07202.html.

Ježek, J., Jedlička, K., Mildorf, T., Kellar, J., Beran, D. (2017). Design and evaluation of WebGL-based heat map visualization for big point data. In *The Rise of Big Spatial Data*, Ivan, I., Singleton, A., Horák, J., Inspektor, T. (eds). Springer, Berlin.

Joliveau, T. (2011). Le géoweb, un nouveau défi pour les bases de données géographiques. *L'Espace géographique*, 40(2), 154–163.

Joliveau, T., Noucher, M., Roche, S. (2013). La cartographie 2.0, vers une approche critique d'un nouveau régime cartographique. *L'information géographique*, 77(4), 29–46.

Kitchin, R., Lauriault, T.P., McArdle, G. (2015). Knowing and governing cities through urban indicators, city benchmarking and real-time dashboards. *Regional Studies, Regional Science*, 2(1), 6–28.

Kraak, M.J. and Ormeling, F. (2010). *Cartography: Visualization of Spatial Data*. Guildford Press, New York.

Krämer, M. and Gutbell, R. (2015). A case study on 3D geospatial applications in the web using state-of-the-art WebGL frameworks. In *Proceedings of the 20th International Conference on 3D Web Technology*, Heraklion, 189–197.

Lambert, N. and Zanin, C. (2016). *Manuel de cartographie : principes, méthodes, applications*. Armand Colin, Paris.

Lienert, C., Jenny, B., Schnabel, O., Hurni, L. (2012). Current trends in vector-based internet mapping: A technical review. In *Online Maps with APIs and WebServices*, Peterson, M.P. (ed.). Springer, Berlin.

Loidl, M., Fischer, F., Traun, C. (2011). Transactional map symbols. At the crossroads of cartography? In *Proceeding of the Geoinformatics Forum Salzburg*. Salzburg.

MacEachren, A.M. and Kraak, M.J. (1997). Exploratory cartographic visualization: Advancing the agenda. *Computers & Geosciences*, 23(4), 335–343.

MacEachren, A.M. and Kraak, M.J. (2001). Research challenges in geovisualization. *Cartography and Geographic Information Science*, 28(1), 3–12.

MacEachren, A.M., Crawford, S., Akella, M., Lengerich, G. (2008). Design and implementation of a model, web-based, GIS-enabled cancer atlas. *The Cartographic Journal*, 45(4), 246–260.

Meier, S. (2016). The marker cluster: A critical analysis and a new approach to a common web-based cartographic interface pattern. *International Journal of Agricultural and Environmental Information Systems*, 7(1), 28–43.

Mericskay, B. (2016). La cartographie à l'heure du géoweb : retour sur les nouveaux modes de représentation spatiale des données numériques. *Cartes et Géomatique*, 229/230, 37–50.

Mericskay, B. (2021). La géovisualisation de données massives sur le Web : entre avancées technologiques et évolutions cartographiques. *Mappemonde*, 131 [Online]. Available at: https://journals.openedition.org/mappemonde/5595.

Mericskay, B. and Roche, S. (2011). Cartographie 2.0 : le grand public, producteur de contenus et de savoirs géographiques avec le web 2.0. *Cybergeo, revue européenne de géographie* [Online]. Available at: http://journals.openedition.org/cybergeo/24710 [Accessed 15 July 2021].

Moser, J. and Koslitz, S. (2016). Pins or points? Challenges in producing cartographically appealing webmaps within an editorial environment for LiD online. In *Progress in Cartography*, Gartner, G., Jobst, M., Huang, H. (eds). Springer, Berlin.

Netek, R., Brus, J., Tomecka, O. (2019). Performance testing on marker clustering and heatmap visualization techniques: A comparative study on JS mapping libraries. *International Journal of Geo-Information*, 8(8), 348.

Netek, R., Masopust, J., Pavlicek, F., Pechanec, V. (2020). Performance testing on vector vs. raster map tiles comparative study on load metrics. *ISPRS International Journal of Geo-Information*, 9(2), 101.

Palsky, G. (2013). Cartographie participative, cartographie indisciplinée. *L'Information géographique*, 77(4), 10–25.

Panchaud, N.H. and Hurni, L. (2019). Smart cartographic background symbolization for map mashups in geoportals: A proof of concept by example of landuse representation. *The Cartographic Journal*, 56(1), 42–58.

Pánek, J. and Burian, J. (2020). Online visualisation. In *Spationomy: Spatial Exploration of Economic Data and Methods of Interdisciplinary Analytics*, Pászto, V., Jürgens, C., Tominc, P., Burian, J. (eds). Springer, Berlin.

Plantin, J.C. (2014). L'avènement de la carte comme médiation. Généalogie des rencontres entre cartographie et théories de l'information. *Questions de communication*, 25, 309–326.

Poidevin, D. (1999). *La carte : moyen d'action*. Ellipses, Paris.

Poorthuis, A., Van der Zee, L., Guo, G., Keong, J.H., Dy, B. (2020). Florence: A web-based grammar of graphics for making maps and learning cartography. *Cartographic Perspectives*, (96), 32–50.

Raposo, P., Touya, G., Bereuter, P. (2020). A change of theme: The role of generalization in thematic mapping. *International Journal of Geo-Information*, 9(6), 371.

Rieder, B. (2009). Visualisation spatiale avec Google Maps. *Hyperurbain II – Nouvelles cartographies, nouvelles villes*.

Robinson, A.C., Demšar, U., Moore, A.B., Buckley, A., Jiang, B., Field, K., Kraak, M.J., Camboim, S.P., Sluter, C.R. (2017). Geospatial big data and cartography: Research challenges and opportunities for making maps that matter. *International Journal of Cartography*, 3(1), 32–60.

Roth, R.E. (2012). Cartographic interaction primitives: Framework and synthesis. *The Cartographic Journal*, 49(4), 376–395.

Roth, R.E. (2013). An empirically-derived taxonomy of interaction primitives for interactive cartography and geovisualization. *IEEE Transactions on Visualization and Computer Graphics*, 19(12), 2356–2365.

Roth, R.E., Donohue, R.G., Sack, C.M., Wallace, T.R., Buckingham, T.M. (2014). A process for keeping pace with evolving web mapping technologies. *Cartographic Perspectives*, 78, 25–52.

Scharl, A. and Tochtermann, K. (2007). *The Geospatial Web: How Geobrowsers, Social Software and the Web 2.0 are Shaping the Network Society*. Springer, Berlin.

Shneiderman B. (1996). The eyes have it: A task by data type taxonomy for information visualizations. In *Proceedings of the IEEE Symposium on Visual Languages*, 336–343.

Seguin, C., Nierobisz, A., Kozlowski, K.P. (2017). Seeing race: Teaching residential segregation with the racial dot map. *Teaching Sociology*, 45(2), 142–151.

Singleton, A. and Brunsdon, C. (2014). Escaping the pushpin paradigm in geographic information science: (Re)presenting national crime data. *Area*, 46(3), 294–304.

Smith, D.A. (2016). Online interactive thematic mapping: Applications and techniques for socio-economic research. *Computers, Environment and Urban Systems*, 57, 106–117.

Sui, D., Elwood, S., Goodchild, M. (2013). *Crowdsourcing Geographic Knowledge: Volunteered Geographic Information in Theory and Practice*. Springer, Berlin.

Tang, W. (2020). Cartographic mapping driven by high-performance computing: A review. In *High Performance Computing for Geospatial Applications*, Tang, W. and Wang, S. (eds). Springer, Berlin.

Tolochko, R.C. (2016). Contemporary professional practices in interactive web map design. Master's Thesis, University of Wisconsin, Madison.

Usher, N. (2020). News cartography and epistemic authority in the era of big data: Journalists as map-makers, map-users, and map-subjects. *New Media & Society*, 22(2), 247–263.

Walker, K.E. (2018). Scaling the interactive dot map. *Cartographica*, 53(3), 171–184.

Wallace, T.A. (2011). A new map sign typology for the GeoWeb. In *Proceedings of the 25th Annual International Cartographic Conference*, Paris.

Ware, C. and Plumlee, M. (2005). 3D geovisualization and the structure of visual space. In *Exploring Geovisualization*, Dykes, J., MacEachren, A., Kraak, J. (eds). Elsevier, Amsterdam.

White, P. and Powell, S. (2019). Code-literacy for GIS librarians: A discussion of languages, use cases, and competencies. *Journal of Map & Geography Libraries*, 15(1), 45–67.

Zanin, C. and Ysebaert, R. (2017). Mise en place d'un modèle cartographique : trouver le bon compromis entre homogénéisation et innovation cartographique. *Cartes et Géomatique*, 233, 33–44.

Zastrow, M. (2015). Data visualization: Science on the map. *Nature*, 519, 402–405.

Zunino, A., Velázquez, G., Celemín, J.P., Mateos, C., Hirsch, M., Rodriguez, J.M. (2020). Evaluating the performance of three popular web mapping libraries: A case study using Argentina's life quality index. *International Journal of Geo-Information*, 9(10), 563.

7

Geovisualization and Cartographic Communication: Style and Interaction

Sidonie CHRISTOPHE
LASTIG, ING, ENSG, Gustave-Eiffel University, Saint-Mandé, France

7.1. Introduction

Geovisualization is the set of knowledge and techniques that allow the visualization of a territory, or a spatiotemporal phenomenon on this territory, by interacting with geographic data and by using the user's perception and cognitive abilities (Christophe 2020). As a scientific field, it feeds on, mobilizes and enriches the principles, methods, techniques and tools of cartography. It participates in the diversification of cartographic communication through interaction with data and (carto)graphic representations, such as diagrams, 3D visualizations or multiple views. Geovisualization makes it possible to revisit the problems of cartographic conception, as a succession of abstractions of spatial reality, with the objective being interactive exploration of the represented data and phenomena.

Historically, and schematically, the field of cartography refers to (often static) spatial representations that facilitate communication, whereas the field of geovisualization concerns visualizations (dynamic and interactive) that facilitate spatial and temporal reasoning, as well as the exploration of complex data (MacEachren and Kraak 2001). However, this interpretation tends to oppose cartography and geovisualization in practice, even if the map allows for spatial reasoning and geovisualization for communication (Christophe 2020). In this

Cartographic Communication,
coordinated by Boris MERICSKAY. © ISTE Ltd 2023.

chapter, we position ourselves on the side of geographic information sciences, fueled by manipulation and cartography, driven by the uses and users of spatial data that are ever more numerous, requiring better modeling of the key concepts of geovisualization, and the design and development of specific tools.

This chapter reviews what geovisualization can bring to (cartographic) communication. The objective is both to present what geovisualization is and to discuss how and why cartographic communication is now largely re-examined by the new uses of digital maps. A return to the historical definition of geovisualization and its evolution in a dynamic context of data, users and varied uses is necessary to better position it. From an overview of the geovisualization tools, we underline the challenges of customization and interaction. We then show how interaction with styles makes it possible to explore more expressive representations by integrating knowledge from cartographic and/or artistic practices. Lastly, we present some applications of geovisualization, on different themes, in order to show how the modes of interaction with styles and data, as well as personalization, favor the appearance of new forms of communication.

7.2. Geovisualization and cartographic communication

The development of digital tools has offered users the possibility to interact with the map and data, thus opening the field to the exploration of cartographic possibilities, beyond the communicative role of the map. Historically speaking, the term *geovisualization* has been defined as "the theory, methods and tools for the exploration, analysis, synthesis and visual presentation of spatial data" (MacEachren and Kraak 2001). The historical goal of geovisualization is to "design and use spatial visual media to explore data, and through that exploration generate hypotheses, develop solutions to problems, and construct knowledge" (Kraak 2003), as well as to help stimulate "individual and collective knowledge about places, distributions, and interactions in space and time" (Dykes et al. 2005). These definitions, which were formalized several years ago, are still valid today, but need to be reformulated in a highly dynamic context of technicality and usage. The repositioning of geovisualization in relation to cartography and data visualization (dataviz) is also necessary in order to better understand its specificity and objectives.

7.2.1. *Geovisualization, a field of research and interdisciplinary practices*

Geovisualization aims to design (carto)graphic representations and means of interaction with geographic data, in order to help a user see, perceive, understand, interpret, analyze and even make decisions about a spatialized phenomenon. This definition requires dealing with visual, semiological and cognitive complexity problems that arise from:

– on the one hand, the complexity of spatiotemporal phenomena to be represented in dynamic interaction with a geographical space characterized by terrain, entities, forms, structures and arrangements, according to what we wish to observe and analyze;

– on the other hand, the capacity of visual integration and interactive exploration of heterogeneous data, according to different points of view and various intentions, triggering mechanisms of perception and cognition that allow a spatiotemporal reasoning specific to the individual (Christophe 2020).

The design of a geographic data visualization system that helps a user (citizen, scientist, decision-maker) perform these tasks remains a complex problem. This design requires knowledge and skills related to the following three objectives:

– to "show" the graphic representation, which is linked to the visual integration (semiotics and information encoding), as well as the graphic rendering of the spatialized data (display of the encoded information on the screen);

– to "analyze", which is related to the interactive exploration of the data, via modes of interaction and adapted interfaces;

– to "give meaning", which is linked to spatiotemporal reasoning, through the activation of cognitive mechanisms (recognition, memorization, learning, etc.) associated with mental representation.

7.2.2. *Challenges on data, user and usage*

Geovisualization and its tools evolve in a dynamic context, in terms of data acquisition techniques and geographic information production processes, in particular collaborative ones, generating more and more spatialized data, as well as in terms of users and uses (expressed, expected, to be invented) of these data through their visualization.

7.2.2.1. *Data and visualization challenges*

There is no shortage of multi-sensor, multi-source, multi-scale, variably accurate and variably massive data. In response to this abundance of information, it is necessary to have solutions to visualize various spatial data (satellite imagery, maps, databases, GPS data, collaborative data, etc.), to add textual archives, geo-historical databases or old photos, to navigate through these data, and to personally add information. Moreover, there is a great expectation to have tools to manipulate, compare, interpret and combine all these heterogeneous data in order to build our own "maps".

There are numerous challenges in spatiotemporal information visualization with respect to acquired, semanticized, simulated and produced multi-dimensional data (2D/3D/nD), as well as collaborative data, which has or has not been geolocalized, and has been created by the users. The main goal is to provide methods and tools for co-visualization (Hoarau and Christophe 2017; Lobo et al. 2017, 2019), interaction and diffusion of repositories and job data, regardless of spatial and temporal scales. This implies integrating various heterogeneities, in source, precision and scale (Nöllenburg 2007), as well as revisiting points of visual, semiological and cognitive legibility and complexity (Fairbairn 2006; Jégou and Deblonde 2012; Touya et al. 2016). The links between abstraction, perception and cognition are central to these considerations.

On the one hand, the tool designer must master the construction of a geovisualization, so that it meets the needs, preferences and usage of the intended users. On the other hand, there must be easy access to the settings of (cartographic) representation methods, depending on the level of expertise of the designer. The tools do not regulate the quality of the design or the final result, often providing predefined representation styles, and there are no means of self-assessment of the semiological fundamentals (Bertin 1967; Morrison 1974; MacEachren 1995).

7.2.2.2. *User adaptation challenges*

A whole field of research exists on the functional use of geovisualizations, their usability, the user experience that is at stake, and on the associated perceptual and spatial cognition challenges (Slocum et al. 2001; Fabrikant and Lobben 2009; Davies et al. 2015)[1]. Facilitating "visuo-spatial" reasoning (Tversky 2005) remains a challenge. A visualization must support a range of user interaction tasks, which may vary according to need and use (MacEachren 1994; Roth 2012). This challenge is always present and transforms as the diversity and quantity of data, users and new visualization techniques increase (Griffin and Fabrikant 2012).

This requires experimenting and assessing the quality and usability of the tools with users (Ooms et al. 2015). A (well)-controlled experiment implies the definition of simple tasks in a simplified geographical space in order to be able to validate

1 For many years, two commissions of the International Cartographic Association (ICA) have focused their work on these topics: one on the cognitive challenges of geovisualization and another on design and user experience.

hypotheses. However, traditional geovisualization tasks are complex and performed in a complex geographical space, not facilitating the definition of independent assumptions, the design of controlled experiments and the reproducibility of user experiments on the choice of colors or a visualization mode (Christophe 2020). This currently remains a major obstacle and does not help to define usable rules about how to build an adequate (carto)graphic representation (Roth et al. 2017).

For the designer, it is important to return to a user-centered design. The notion of "context of use" is interesting because it encompasses all relevant information that describes the user and their environment (Dey 2001), their activities during use, the temporal conditions and physical environment, and the system and technology behind the application (Griffin et al. 2019). Better modeling the context of use allows for the extraction of rules for graphical design. The current approach is to consider that there are more representations that simplify a cartographic task, in a given context, than a single solution (Griffin et al. 2019). For any given user, it provides a framework for modeling, development, and to design geovisualization tools that are adapted, usable and above all scalable.

7.2.2.3. *Challenges of adapting to uses*

Geovisualization tools must therefore adapt to uses that have diversified over time, and that now refer to multiple contexts of use and users with varying levels of expertise. This adaptation to uses, which is a particularly complex and delicate subject (Blaise et al. 2015; Ritschard et al. 2016; Bonaccorsi et al. 2018), refers to the very designers of the tools. Designing a geovisualization tool can only be completed with the intended users, based on usage that has to be understood, from the initial stages of the design process, and throughout the process (Lloyd and Dykes 2011). Designers must anticipate and identify (future) uses, in order to propose functionalities and modalities that are in line with practices. This implies adopting methods for collecting user needs, which involves experts, specialists in field surveys, collection techniques and actor interactions, in order to be able to translate the observed uses into methods and to implement functionalities, while considering the limitations encountered and the feedback from experience.

7.2.3. *Linkages with cartography and data visualization*

The design process in geovisualization partly takes up the definition of a map proposed by Palsky (2021), "a mental image that relates to something, and the notion of the process by which it is created". However, unlike a "traditional" approach, where the cartographer produces a map that is often static, with a choice of data that is already shaped, a defined style, etc., in geovisualization, each of the steps in the construction of the map is interactive and dynamic, focused on the user's side (data selection, geometric preprocessing, generalization, symbolization, etc.).

While the map presents the result of an analysis to be communicated, geovisualization provides a set of tools for interacting with the data, their style of representation and the views proposed, and ultimately for constructing one's own maps. The logic and modalities of cartographic communication are therefore quite different in cartography and geovisualization, insofar as the former refers to a rather linear approach (sender-receiver), and the latter is more in line with an iterative, exploratory and multi-dimensional approach. This multi-dimensional aspect can increase, depending on the spatiotemporal scales, the dimensions (2D, 3D, other geometries) and the data sources. This is an significant development with regard to (carto)graphic communication.

Geovisualization partly overlaps with the objectives of information visualization (Card et al. 1999; Kerren et al. 2008) and data visualization (Post et al. 2003; Friendly 2008). The convergence between these fields is strong when it comes to spatializing data on a base map. The boundary between map, data visualization and geovisualization is becoming increasingly blurred in new uses, especially on the web. This "mix of genres", where diversified profiles and heterogeneous practices intermingle, largely questions the cartographic communication process[2].

7.3. Geovisualization tools, between personalization and interaction

From examples of geovisualization tools, we specify how the personalization of maps is proposed to users, and how it allows them to visually explore the data. This panorama of tools leads us to highlight the difficult challenges for rethinking interaction in geovisualization.

7.3.1. *Map customization and visual data mining*

7.3.1.1. *Map personalization*

The involvement of user-designers has become important in personalized cartographic production. Access to reusable or even modifiable base maps and to open geographic or cartographic databases, via data visualization interfaces (OpenStreetMap, in particular), has opened new perspectives. The customization of base maps, or of certain elementary data, is an integral part of the problems of geovisualization. Numerous tools have gradually emerged to provide base maps

2 This cross-cutting issue in the book is addressed extensively in Chapters 2 and 6.

with various styles (e.g. Google Maps styling tool and MapBox). Figure 7.1 illustrates how, in advanced mode, it is possible to specify the style for each data layer, so as to create a customized base map that can be reused in a geovisualization application, an online map or using GIS software.

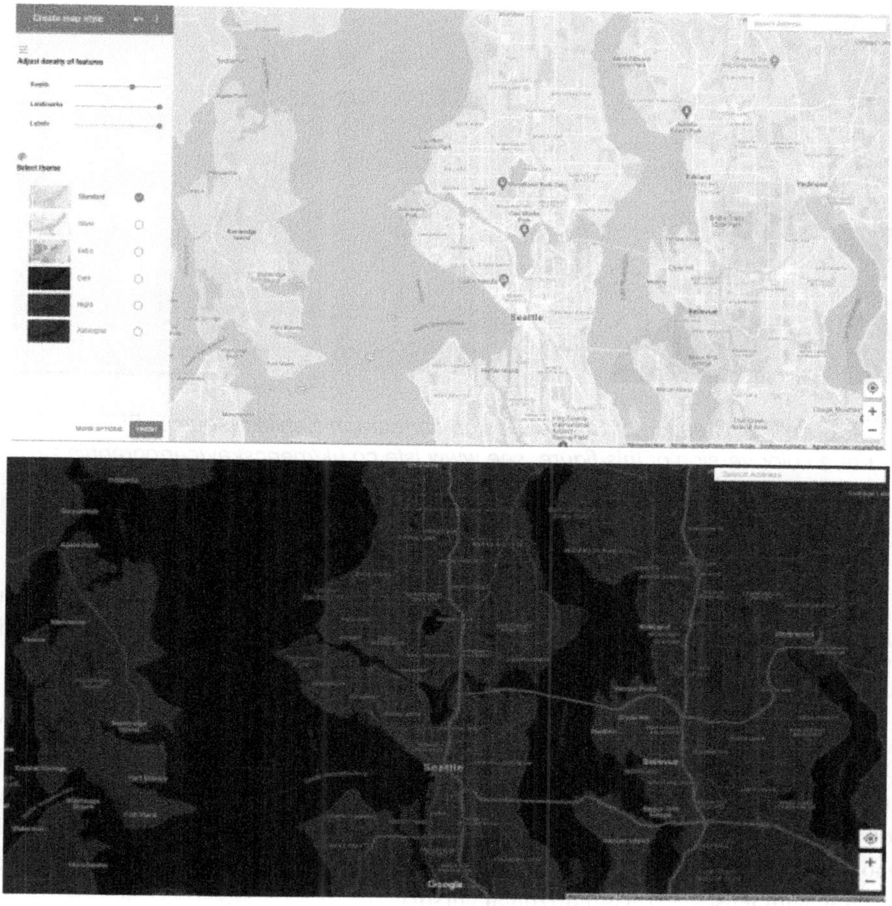

Figure 7.1. *Base map styles (source: Google Maps styling tool). For a color version of this figure, see www.iste.co.uk/mericskay/cartographic.zip*

Figure 7.2 shows how to overlay and combine up to five base map styles to obtain a new base map style, by playing with transparency, opacity and various blending techniques (blending modes) from image processing (multiplication, addition, differentiation, filtering, avoidance, etc.). Other more or less expert tools make it possible to modify the style of the base maps, the geometry and the semiology of the elementary data, from palettes, signs or styles that are predefined or to be created[3].

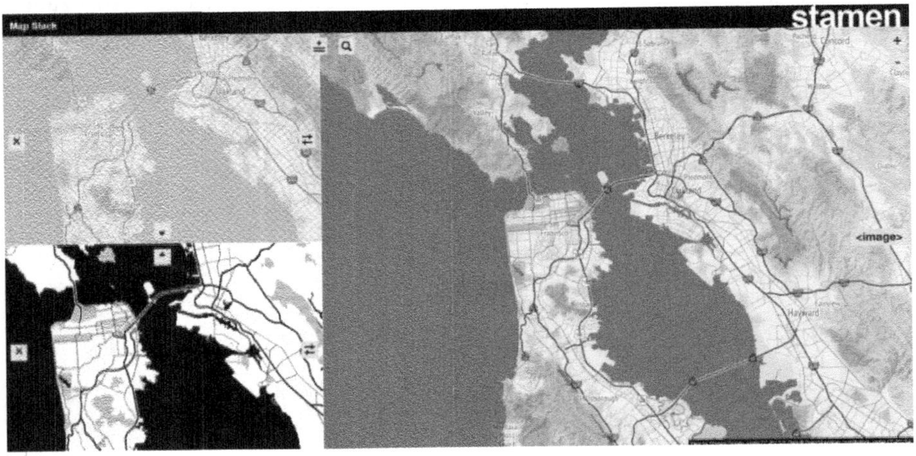

Figure 7.2. *Creation of StamenDesign cartographic styles (source: Stamen). For a color version of this figure, see www.iste.co.uk/mericskay/cartographic.zip*

7.3.1.2. *Exploring data*

Another form of personalization consists of allowing the user to select data and/or locations, to represent them graphically, and to observe evolutions and changes in space and in time.

The ability to manipulate images of the territory (old, historical, postcards, etc.) and maps has become important, with a view to enhancing the value of historical and patrimonial analysis of territories. The IGN's "Go back in time" tool (Figure 7.3), for example, makes it possible to co-visualize old and current maps and aerial photos within the same interface, via different synchronized interaction modes (side by side, with lenses, vertical/horizontal cursor).

Numerous online mapping applications make it possible to manipulate old images from iconographic collections, which have been made available online by national archives or museums, or from collaborative initiatives (photos, postcards, etc.). The Historypin project, for example, geolocates a set of historical photographs, by pointers on a base map, allowing access to collections of images according to a location or a keyword. Side by side, the user can visualize a map referencing the locations where he can find images, and the corresponding image collection

3 For more on this topic, see Chapter 6.

(Figure 7.4). The Smapshot and Photo Tourism applications allow the user to explore historical images that have been co-visualized directly in a 3D environment, representing the current territory, while leaving a great deal of interactivity to the user. Numerous collaborative research projects are working to improve the georeferencing and geovisualization of historical or heritage images, allowing archivists, historians, archaeologists, planners and the general public to explore this rich heritage. The digital reconstruction of the Notre-Dame Cathedral is interesting in this respect, insofar as it aims to bring together a large number of professions and knowledge around a historical and contemporary (geo)visualization of the building[4].

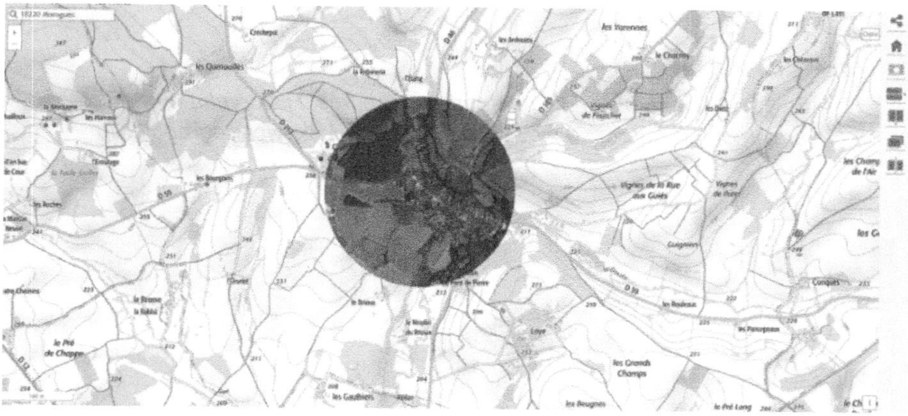

Figure 7.3. *Co-visualization of maps and aerial imagery in the "Go back in time" application (source: IGN). For a color version of this figure, see www.iste.co.uk/mericskay/cartographic.zip*

Creating, sharing and enriching knowledge about the dynamics of spatiotemporal phenomena requires visual approaches to spatial analysis. Geovisualization tools (Andrienko and Andrienko 2006; Kraak 2008) have long incorporated the principles of time-geography (Hägerstrand 1970; Saint-Julien 2001) in order to make sense of visual reasoning about spatiotemporal phenomena. Geovisualization tools effectively use and revisit the principle of multi-view interfaces, each representing an aspect or a point of view of the problem (spatial, temporal, thematic), while all being synchronized.

4 CNRS/French Ministry of Culture site for the restoration of Notre-Dame de Paris.

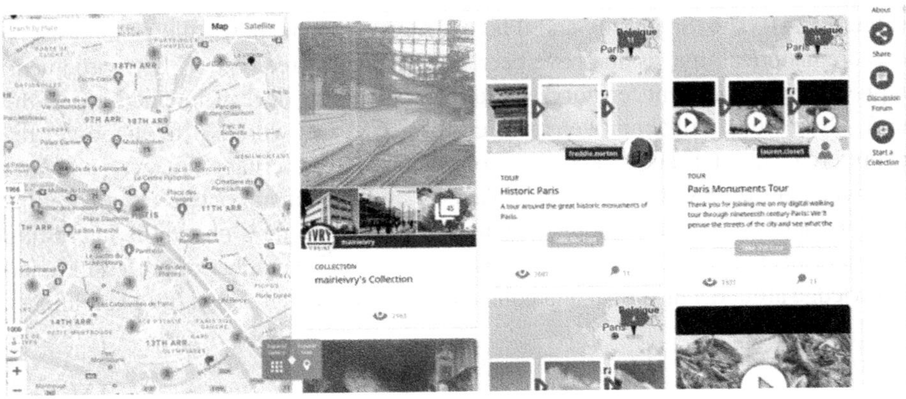

Figure 7.4. *Viewing historical photographs (source: Historypin). For a color version of this figure, see www.iste.co.uk/mericskay/cartographic.zip*

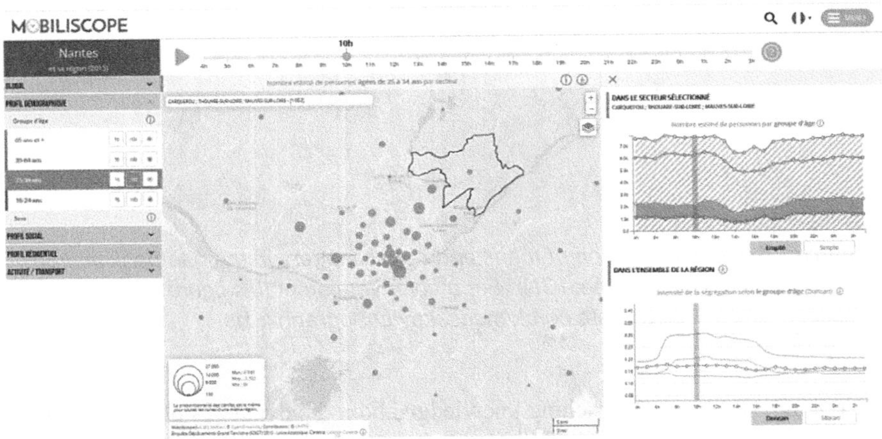

Figure 7.5. *Mobiliscope interface (source: the Mobiliscope). For a color version of this figure, see www.iste.co.uk/mericskay/cartographic.zip*

For example, Mobiliscope allows users to interactively explore the present population and the social mix over 24 hours, for more than 50 cities around the world (Lecomte et al. 2018). This tool, composed of a dynamic map, a temporal scale, and a set of graphs, offers the user a rich interface, using a dashboard, allowing them to navigate through complex data from travel surveys (Figure 7.5). The application is intended for use by planners to assist in decision-making, researchers as an analytical tool and the general public as an educational tool. Beyond a geovisualization interface, it should be noted that the development of

Mobiliscope is in line with the FAIR principles; in other words, the project data are easy to find, accessible, interoperable and reusable, in a logic of reproducibility of the research.

The GrAPHIST geovisualization tool (geovisualization for the analysis of spatiotemporal phenomena) aims to integrate the cyclical dimension of the series of spatiotemporal events (historical, epidemiological, meteorological, social), by providing a dynamic and interactive map, as well as graphs, within the same interface (Figure 7.6). In particular, this web application facilitates the identification of patterns in spatiotemporal distributions, by playing on the semiotic representation of cyclic time and on the way to manipulate the temporal scale (Gautier et al. 2017).

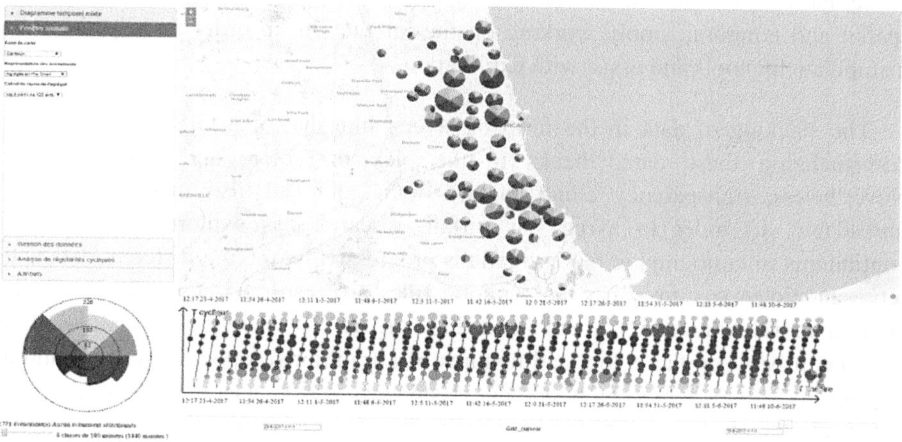

Figure 7.6. *GrAPHIST application interface (Gautier et al. 2017). For a color version of this figure, see www.iste.co.uk/mericskay/cartographic.zip*

These tools offer multiple combined interfaces with thematic maps, diagrams and graphs that are all synchronized with each other, with which it is possible to interact in order to help explore the represented data.

7.3.2. *Challenges around interaction with data*

Navigating and exploring heterogeneous spatial data sets implies that the user can display, superimpose and combine different layers of information, change the representation modes, change the perspective and move from one dimension to another. This interaction between the user, the data and the geovisualization must be done without loss of visual reference points and implies that the user, who is not necessarily a specialist or someone used to these tools, is at least guided.

The complexity of the data and the spatiotemporal phenomena requires that the geovisualization tools proposed allow the user to change his/her point of view. On the one hand, the latter refers to the place from which the user observes the represented space (real geographical position and point of view in the representation) through a digital device (screen, camera, 3D camera, augmented reality glasses). On the other hand, the user can visualize the spatial data under multiple angles (top view, oblique view, while immersed, pedestrian height, roofs, etc.). He can also navigate higher or lower in altitude, deeper or lower in depth in a scene, more or less far into the past, or into the future. All these dimensions, both spatial and temporal, imply working on the interaction, in order to help the user manipulate the tools and assist with exploration.

The stacking of data in the form of layers, like that of a GIS, contributes to co-visualizing these data, thanks to the play on colors and transparencies. Nevertheless, transparency alone has its limits and can also make everything unreadable. In order to avoid this pitfall, research has explored the idea of continuums of cartographic representations progressively, mixing different data and representations, ensuring that the semiotic rules are respected at any point of the continuum (Hoarau and Christophe 2017). Figure 7.7 shows an example of a continuum between the colors and textures of a traditional topographic map, and a Pop Art style map, with a satellite image. Using a slider, the user can introduce more or less photorealism into the map, or more or less abstraction into the image, and can thus choose the best mix of colors and textures according to their needs.

It is also possible to work on the notion of interface with the user. The multi-view mode remains an efficient process, where it is possible to co-visualize different parts of the territory at different spatial scales or containing distinct information. Multiplexing techniques, consisting of displaying data "simultaneously" by juxtaposition (such as multi-view) or superposition (such as lenses), have been widely explored and assessed in the field of human–computer interactions (Pietriga et al. 2010; Pindat et al. 2012; Lobo et al. 2017). The design of intuitive and innovative interfaces, facilitating interaction and data exploration in order to perceive space in a possibly different way (Christophe 2020; Çöltekin et al. 2020), is an important issue. Depending on the context of final use, the user must be able to benefit from these changes of perspective through a possible interaction with the visualized scene, and also be able to switch to 2D or 3D, or even dive into immersive environments, increasing their experience with the data and the phenomena represented.

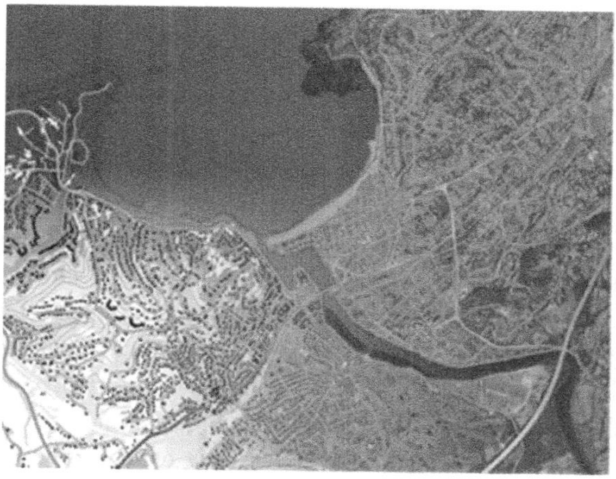

Figure 7.7. *Continuum of topographic styles between IGN Classic Top25, Pop Art map and satellite image (Hoarau and Christophe 2017). For a color version of this figure, see www.iste.co.uk/mericskay/cartographic.zip*

7.4. Interacting with style for more cartographic expressiveness

Interacting with the manner in which data are graphically represented is a way to offer more exploration through customization. It is important to not only look at the set of visual variables available, but also at more complex visual properties, in order to control the quality of a graphical representation. With Geoweb technologies, the availability of new spatial data, and a set of tools dedicated to customizing base maps (Mericskay 2021), the notions of style and stylization have become fashionable in cartography (Figure 7.8). This approach to the search for expressiveness through style, combining creativity and quality, remains an important research topic, to the extent that this new form of map design has largely left the cartographers' box and holds a significant place in geovisualization practices.

Figure 7.8. *Variety of StamenDesign cartographic styles (source: Stamen). For a color version of this figure, see www.iste.co.uk/mericskay/cartographic.zip*

7.4.1. *Back to the notion of style*

The notion of style refers to a set of characteristics and thus recognizable practices in the way something is designed or done. In architecture, for example, style refers to ways of building, using certain materials, forms and structures. In computer graphics, which is a science that involves creation, manipulation and rendering of digital images, the style of an image is analyzed based on knowledge of stylistics, a branch of general linguistics by Willats and Durand (2005). Most individuals are able to recognize stylistic similarities and differences between different schools of painting or artists' works. The analogy with language suggests that (being able) to recognize these differences may depend on an implicit knowledge of the graphic rules, such as rules of grammar that generated these images (Willats and Durand 2005). The underlying challenge, which would allow for a better exploration and control of the expressiveness of style, is figuring out how to make these rules explicit (Christophe 2020).

Today, there are several sets of rules that provide a framework for the cartographic representation of data. These include the grammar of graphics for data visualization (Wilkinson 2005), graphic semiology for cartography (Bertin 1967; MacEachren 1994; Morrison 1994) or proposed syntax rules for online mapping (Mericskay 2016; Poorthuis et al. 2020). Nevertheless, these rules do not describe how to achieve a global composition to obtain a map style, in the way a painter creates a painting from his own "rules of composition" whose style can be recognized (personal style of the painter, style of a school, style of an era, etc.). In cartography, sets of rules and properties on visual variables guide the design, but do not establish the recipe of the cartographic composition, which will yield the final style.

7.4.2. *Style in cartography, two levels of definitions*

The notion of style refers to a set of design practices, which are visually recognizable. It refers to the way in which elemental objects are represented and with which tools they have been elaborated, and to the final pictorial composition. Style in cartography refers to two levels of definition:

– the clarification of the rules for generating graphic objects, specifying how geometries should be symbolized (e.g. a linear length with a thickness and a color), and how this style should be applied by the rendering engine on the screen (graphic rendering), for a range of visualization scales;

– the composition of a final image conveying a style (e.g. an old style, a cubist style, a cartographer's style), made of recognizable characteristics, in the choice of colors, shapes, pencil strokes, visually prominent objects, the way vegetation or waterways are represented, etc.

7.4.2.1. *Specification of the rendering of symbolized geometries and the blending of layers*

In online cartography as in geovisualization, style has been an integrated notion for many years, notably through its technical formalization by the open geospatial consortium. As part of its mission to standardize geographic information, this organization has produced two standards for the graphical representation of spatial data, which are now implemented in most online services and GIS software. The Symbology Encoding (SE) standard describes the language for defining a style, while the Styled Layer Descriptor (SLD) describes the application of a style to a geographic data layer (Mueller 2006; Lupp 2007; Ertz et al. 2020).

In this context, style is a technical specification of the representation of each layer of typical geometric objects (point, line, polygon), on which a composition of color, thickness and type is applied and adapted to the geometry concerned. This assumes that the rendering engine of the cartographic tool is capable of effectively rendering the digital display. In existing cartographic design tools or GIS, the description of a style is done when selecting a sign for a data layer, and is saved in a style file, layer by layer, or for the whole set of layers. The set of styles is specified in a configuration file, which can be read by the graphics rendering engine of the cartography/visualization/GIS tools.

These tools leave the user completely free to define elementary styles, and to obtain a final map style, without any other form of validity or quality (legibility, efficiency, usability) than the user's own choice. Nevertheless, the art of constructing a legend, which is seen as a set of constraints on these styles, so as to obtain an efficient map, remains the responsibility of the user; the legend "controls" the quality of the assignment of graphic signs according to current constraints, between the themes and intrinsic themes of the legend, and the user must have either the expertise to preserve its quality or the tools to help them do so (Christophe 2011).

7.4.2.2. *Overall pictorial style: salient and recognizable characteristics*

Style also refers to a second, broader definition, that is, a general impression reflected by the (carto)graphic representation, composed of salient and recognizable characteristics. For a painting, this can be a color palette, a particular color, a contrast, the movement of characteristic brushstrokes, a play of light, etc. For a map, it may be a color palette, a way of representing waterways and buildings, vegetation areas, a relief, place names or a typography.

Cartographic styles are a specific way of showing geographic information, based on particular history and practices, partly related to different ways of using "content and appearance" (Kent and Vujakovic 2009; Beconyte 2011; Christophe 2012). Appearance depends on choices of graphic representation using visual variables and their possible combinations, toponymy and graphic abstraction. These characteristics can identify a time period, a movement, a cartographer or an institution. This definition of style refers to a function that allows a map to belong to a representative family of maps and types of legends.

7.4.3. *The example of topographic map styles*

Identifying and analyzing existing topographic styles allows us to understand what constitutes a map family and to work on its evolution, by integrating fine-tuned modifications, without disturbing user reading habits. Map producers, whether they are institutional or not, have their own graphic codes, which their users have consciously or unconsciously integrated.

When a person that is used to using IGN Top25 maps for hiking suddenly discovers a 1:25,000 hiking map from another institute, there is an inevitable period of adjustment. If students are asked to represent a default data set, the majority of those who have lived in France will make style choices mixing IGN and Michelin styles, keeping color and legend options very close to the maps they have been accustomed to seeing and consulting. When Google reintegrates country-specific representations of public transportation symbols, it is to adapt to the user's understanding and habits to facilitate visual location.

Mountain maps convey styles that are specific to cartographic practices, such as the drawing of rock in Swiss topographic maps (Jenny and Hurni 2006; Geisthovel and Hurni 2018). Depending on the data and scales of representation, these cartographic styles are part of the historical evolution of practices, technical advances and user uses and needs that have changed over time. Figure 7.9 illustrates the idea that it is possible to associate a specific topographic style with a country.

In order to better understand these visual, perceptual and cognitive mechanisms, it is important to identify this graphic code, beyond the legend. Understanding what makes a user recognize a map, or a legend, will eventually allow us to reuse this visual code and to guide user attention, especially in the exploration of spatial data in cross-border contexts. Figure 7.10 shows the comparison of the topographic representation of the city of Thonon in the IGN style and the Swisstopo style. These two map styles are very different and characteristic of the practices, conventions, histories and uses of topographic maps in these countries.

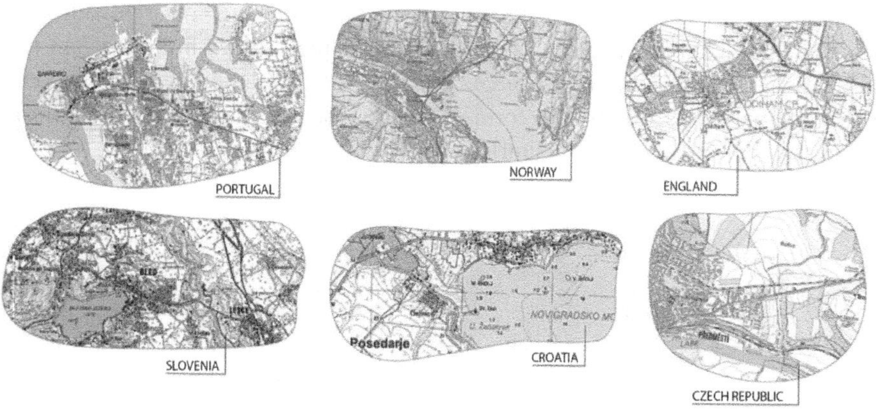

Figure 7.9. *Variety of topographic styles in Europe (Ory 2017). For a color version of this figure, see www.iste.co.uk/mericskay/cartographic.zip*

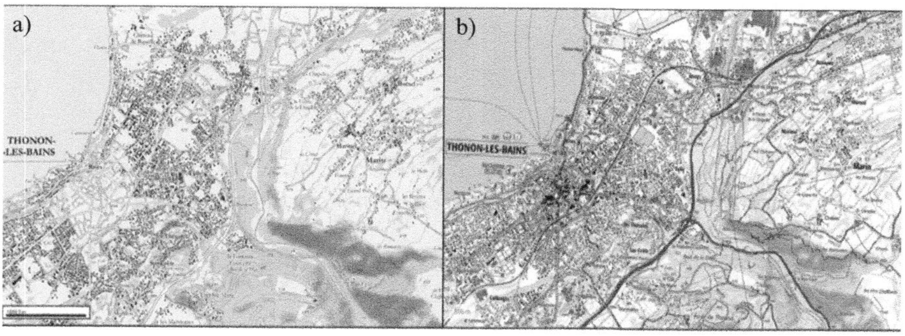

Figure 7.10. *Comparison of topographic styles: (a) Swisstopo and (b) IGN (Ory et al. 2018). For a color version of this figure, see www.iste.co.uk/mericskay/cartographic.zip*

Identifying visual signatures that are characteristic of these styles is the subject of further research; user experiments have focused on the recognition of visual characteristics, what the user actively identifies as representative of a topographic style and what he remembers as representative marks, from a set of maps. Graphical signs representing tourist information, shading and contour lines are highly significant in recognizing an IGN map, while those representing toponymy, buildings, vegetation and the road network are highly revealing in recognizing a Swisstopo map (Ory et al. 2015, 2018). Formalizing these visual signatures paves the way for reusing some of their major features, as decoding pivots, to build map continuums. For example, a mix of Swisstopo and IGN topographic styles allows for the creation of a hybrid style of the French-Swiss border, shown in Figure 7.11 (Ory et al. 2018), which can be reused in Geoportals. This facilitation of visual decoding of topographic styles results in a form of spatial visual continuity, when exploring the territory between different countries.

Figure 7.11. *Mix of Swisstopo and IGN (Ory et al. 2018). For a color version of this figure, see www.iste.co.uk/mericskay/cartographic.zip*

7.4.4. *Pictorial, graphic and artistic styles*

Identifying pictorial, graphic and artistic styles that can be transferred to the map, especially in geovisualization environments, responds to a need for expressiveness of the cartographic message. It is possible that a visual variable adapted to the information to be represented is missing (e.g. the uncertainty of data, a spatial phenomenon or a particular relief), or that one wants to reproduce a line drawing, a watercolor, a line with a progressively evolving thickness, or the typical contrasts of Fauvism. Tools exist to manipulate styles that have already been predefined, such as watercolor or an old style. However, it is difficult to get out of default styles, especially if the user is aiming at a specific pictorial style, or simply wants to introduce something more creative and artistic.

7.4.4.1. *Color styles*

A first approach consists of exploring the space of possibilities in terms of colors to manipulate and transfer it into the map. The idea here is to not only think about the choice of colors, but to also consider how to use them according to the cartographic rules and conventions on colors, or according to one's own context of use, needs and preferences. A large literature exists on this issue (Brewer 1992; Harrower and Brewer 2003; Jégou 2018; Christophe 2019), and many tools allow the creation and manipulation of color palettes.

If the user wishes to be inspired by pictorial and artistic styles, the underlying question is how does one transfer these colors and the associated pictorial composition rules to the space of the (carto)graphic representation? Research studies have made it possible to characterize these constraints (mapping rules, compositional rules of the painter, user preferences) through dialogue with the user, in order to co-construct palettes adapted to their data set, by selecting colors from maps and/or paintings, according to their preferences (Christophe 2011).

The exploration of color palettes has made it possible to use colors that are different from what we are used to seeing, which can be used to convey a particular message, and that are sometimes surprising, even confusing, with respect to traditional topographic cartography. The specification of a Pop Art style, largely inspired by the works of the artist Roy Lichtenstein, has, for example, led to the creation of online base maps based on these graphic codes (Christophe and Hoarau 2012). In the academic sphere, this approach was then widely taken up by private online map publishers like ESRI or Mapbox. Extracting and exploring color palettes suitable for mapping are actions that then become very important for generating original palettes, while taking into account the constraints suitable for these colors, with the user then having a strong point of view on controlling the exploration of possible color palettes (Mellado et al. 2017).

It is interesting to note that the preferred color palettes at the beginning of this work in 2006 were still quite conventional, especially for topographic maps. Going beyond conventional color palettes was indeed considered "eccentric", especially in topographic mapping (even though European topographic maps sometimes showed radically different and very colorful color choices). Originality at that time was in the inversion of the main color contrasts. For example, dark mode (map on a black background) has been used for a long time to compensate for night vision or darkness, in maritime, road or aeronautical navigation. However, it was not really accepted or used in cartography in other contexts. Today, however, this style of map has become very popular in the field of data visualization and online mapping. In particular, it allows you to easily visualize a data set rendered in tinted and bright colors, as a very efficient support for contrasts. These temporal changes on the acceptance of original colors for cartography show how cultural aspects, habits, conventions and practices play a strong role in the appropriation and acceptance of a new map style.

7.4.4.2. *Texture styles*

Beyond color, the style of spatial entities or a map is also expressed by the strokes and textures used. Based on the knowledge and practices of technical and artistic drawing of old maps, it is possible to describe different types of expressive textures that can be used to fill surfaces in a map in an automatic and customizable way. The MapStyle[5] research project, for example, has made it possible to specify and implement styles of old or hand-drawn textures within digital maps. The reintroduction of textures into the map leads to a better perception of land use (forests, bushes), reliefs (rocky areas, screes) and the sea; the manipulation of textures allows the use of different types of strokes and the application of watercolor effects (Christophe et al. 2016), thus contributing to the design of new styles for cartographic communication. The texture generation proposed by Loi et al. (2017) has therefore made it possible to propose a first approach to the generation of texture sets, which can be reused in thematic mapping, and was used to produce the texture of rock in the mountains, as presented in Figure 7.12.

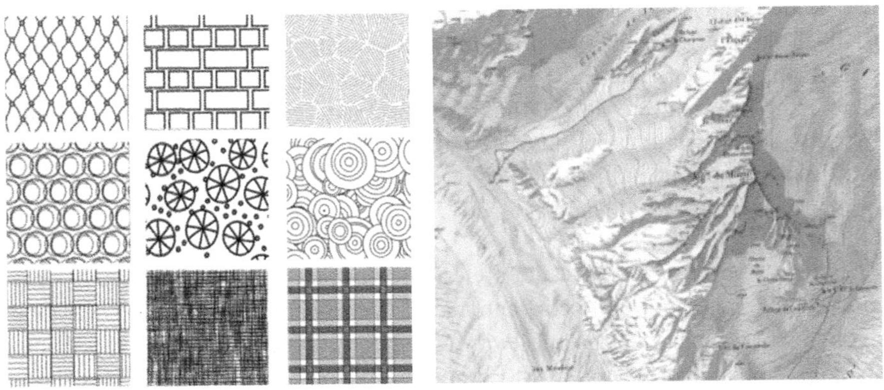

Figure 7.12. *Texture generation, ANR MapStyle project (Loi et al. 2017). For a color version of this figure, see www.iste.co.uk/mericskay/cartographic.zip*

This research has extended the topic of textures within broader practices and wider audiences through the implementation of semiotics and style specifications

5 The objective of the ANR MapStyle project was to study methods from computer graphics to help stylize cartographic renderings, while taking the specificities and constraints of cartography into account (legibility, efficiency, esthetics). It gathered researchers in geovisualization and expressive rendering in a multidisciplinary approach, imitating or inspired by traditional media.

within OGC standards (SLD and SE) (Christophe et al. 2016; Duménieu and Christophe 2016). These approaches allow for the specification and generation of various styles (watercolor, brushstrokes, Cassini map), be it elementally (by object type) or comprehensively (for all objects in the map) (Figure 7.13). Beyond a predefined and fixed style, the user here can modify the level of abstraction, as well as the level of combination in a visualization. This setting and this possible control over the user on the style make it possible to refine the communication potential of these styles.

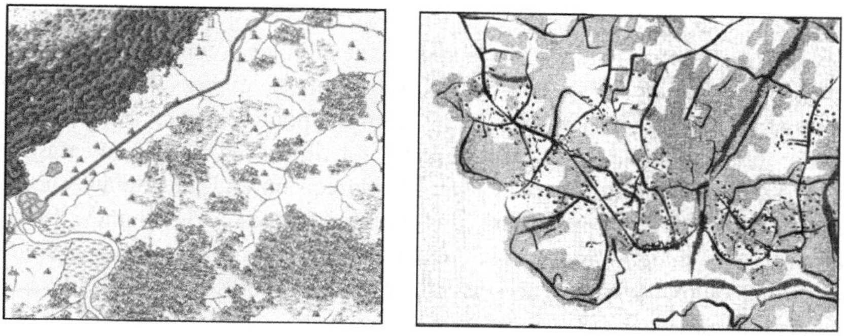

Figure 7.13. *Generation of expressive styles (textures, line drawing, watercolor) for (carto)graphic representation (Christophe et al. 2016). For a color version of this figure, see www.iste.co.uk/mericskay/cartographic.zip*

To improve the quality of (carto)graphic representations and, in particular, their role in the communication aspect of geovisualization, it is important to pursue a better knowledge and modeling of the space/style/message system, as well as existing communication models.

7.5. Diversification of representation modes and interaction for new forms of cartographic communication

Geovisualization allows us to take another look at the data we manipulate and represent. The latter can be observed "in another way", with "another point of view". Making the most of the exploration aspects of available data is not only a way to show and analyze the latter differently, but also to communicate in another manner. Indeed, to transmit complex information, for equally complex uses as a communication medium, mediation or decision-making, the manipulation of visual variables alone can be insufficient and limit the possibilities.

Being able to explore the complexity of a set of objects or a phenomenon, thanks to the manipulation of the possible levels of abstraction (conceptual, geometric, graphical, mental), remains a challenge. In order to illustrate and exemplify all these approaches, this last section will present several geovisualization applications that have been designed and developed with end-users, according to their specific contexts of use[6].

7.5.1. *Interaction with style to help visual focus*

3D geovisualization, which lacks a specific semiological conceptual framework, benefits from the revisiting and generation of more expressive (carto)graphic representation styles. Rather than being entirely photorealistic, integrating and exploring (carto)graphic abstraction allows for improved visual attention and collective reasoning, with a certain level of imprecision and uncertainty about the objects being discussed, particularly in terms of shapes and volumes. In some contexts, like planning, introducing too much detail is not necessary, being aware of volumes and empty or full spaces is important, and perceiving possibilities and potentials is an objective.

> Citizen consultation in the development of public policies is often based on 3D digital models that allow for joint reflection on urban planning. Even though it suitable when showing and explaining phenomena such as flood risks in the city, the use of photorealistic 3D visualizations is limited when it comes to questioning the evolution of building forms in neighborhoods (Jacquinod et al. 2016). A certain level of graphical abstraction is required to guide visual attention to the forms and volumes to be built, in empty spaces of the city, interacting with solid spaces. The interactive exploration of building projects based on abstract, expressive and even artistic (in the sense of the line used) styles, allows urban planning agencies to build models that better guide consultation, and keep an "aspiring" mindset (Brasebin et al. 2016). Figure 7.14 shows possible building styles (typical, draft, transparent) on which anything can be set up; these styles can be applied to the forms to be built (visual focus area) and to the surrounding buildings (visual context), as in the two examples shown. The capacities for abstraction and expressiveness were appreciated and allowed agencies to create their own styles of representation (according to their own needs, preferences, knowledge and expertise of their context of use) to use during a consultation.

Box 7.1. *Abstractions for citizen consultations*

6 All these geovisualization applications are part of research projects involving academic, operational and citizen actors.

Figure 7.14. *Interactive exploration of styles for 3D (Brasebin et al. 2016). For a color version of this figure, see www.iste.co.uk/mericskay/cartographic.zip*

7.5.2. *Interaction with styles for more realism*

The realism of a scene or a spatiotemporal phenomenon does not just come from high resolution images, facilitating the reconstruction of a photorealistic scene. Acting on the levels of graphical abstraction also makes it possible to represent the realism of a localized phenomenon differently.

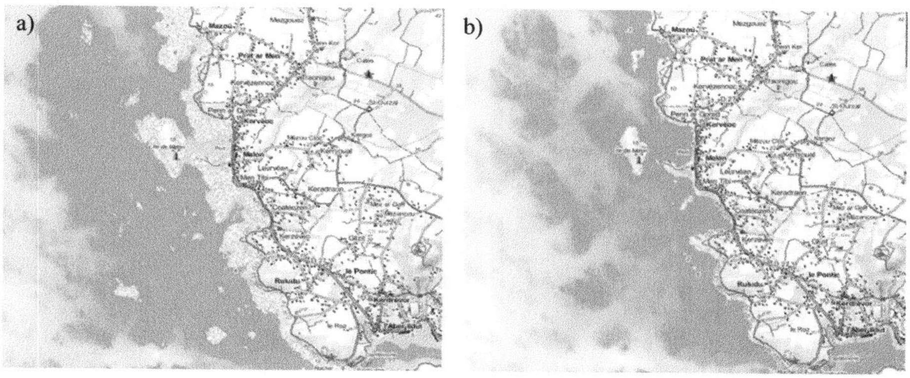

Figure 7.15. *Low tide (a) and high tide (b) in an abstract style, which can be animated over time (Masse and Christophe 2016). For a color version of this figure, see www.iste.co.uk/mericskay/cartographic.zip*

The cartographic data used to represent the coastline often come from different producers, whose contexts of use and practices are different. The combination of these maps in the same geovisualization, such as on the coastal zone, sometimes gives results that are difficult to read and to use, because of the very heterogeneous levels of detail, the represented objects and the styles. One of the tracks consists of making the styles of the represented data more homogeneous, in order to answer to general public uses, for example, to visualize the height of the tide on a certain day at a certain time, to plan a hiking path, or to know if a rocky island is accessible by foot or not. Color palettes applied to the water height changing in time lead to the implementation of temporal animations of the tide, with a traditional topographic style (Figure 7.15). This type of animated geovisualization allows for the appearance of emerged/submerged rocks, discontinuities in possible hiking paths, the height of the tide on the beach, etc. (Masse and Christophe 2016).

Box 7.2. *Temporal animation of the tide, using an abstract style*

7.5.3. *2D and 3D interactions to multiply points of view*

Interaction with data is necessarily dependent on the object to which it is applied, as well as on the level of expertise of the user and his/her mastery of the spatiotemporal phenomenon represented or the manipulation of spatial data. The phenomena of rising tides (including the previously shown tide) present complex questions for geovisualization. Switching to the third dimension to reveal a localized phenomenon is particularly relevant in order to better understand the spatial interactions between such a geophysical phenomenon, its extent and its impacts (buildings and land use).

Figure 7.16. *Visualizations of rising water scenarios (source: iTowns) (created by LASTIG/GEOVIS team, Antoine Masse, Imran Lokhat). For a color version of this figure, see www.iste.co.uk/mericskay/cartographic.zip*

The phenomena of marine submersion, flooding risks or prospective scenarios of rising water levels need to be analyzed in interaction with the geographic space concerned, which is simulated in order to identify behaviors and impacts, and communicated to experts for decision-making and to the general public for information. This diversity of complex data usage and challenges requires visualization approaches that facilitate the communication of these hazards and issues, in both 2D or 3D as needed, with a view of comparing scenarios in space and time, and requires advanced geovisualization capabilities (Perrin et al. 2020). The change of point of view and perspective within geovisualization interfaces allows for additional approaches in the communication, analysis and explanation of these phenomena. Figure 7.16 shows different points of view, depending on whether one wants to observe the extent of the flood risk (left) or the impact on buildings (right).

Box 7.3. *Communication on rising water scenarios*

7.5.4. *Visual exploration in spatiotemporal scales*

Modifying the geometry of the manipulated data, playing on the level of geometric detail, or changing the observed dimension through interactions allow expert users to explore complex data.

Urban simulated temperature data makes it possible to analyze and communicate on the spatial and temporal dynamics of urban air temperature, and particularly the urban heat island phenomenon. Meteorologists working on these questions have great needs in terms of spatial visualization of simulated data in order to identify specific interactions between temperature and urban morphology. To this end, it is necessary to be able to observe the simulated temperature information, the spatiotemporal evolution of this temperature in the urban space, and the way in which what is known about the urban space via morphological indicators impacts the evolution of this temperature. In order to represent the air temperature from complex simulation models, different representations can be developed in the 3D environment, in the form of horizontal planes, vertical planes in the middle of the street and a point cloud colored with a standard color palette. Figure 7.17 illustrates how the user can modulate and combine geometries and map views, as well as modify the formatting properties in just a few clicks to geovisualize this type of 3D spatiotemporal data from different perspectives. Temperature data can be filtered in time and space to facilitate visual analysis. It is important to note that the semiological and cognitive complexities are prominent in this geovisualization, but that the choices of color palettes are in line with the usual representational practices of end-users (Gautier et al. 2020).

Box 7.4. *Interactive visual exploration to understand urban heat islands*

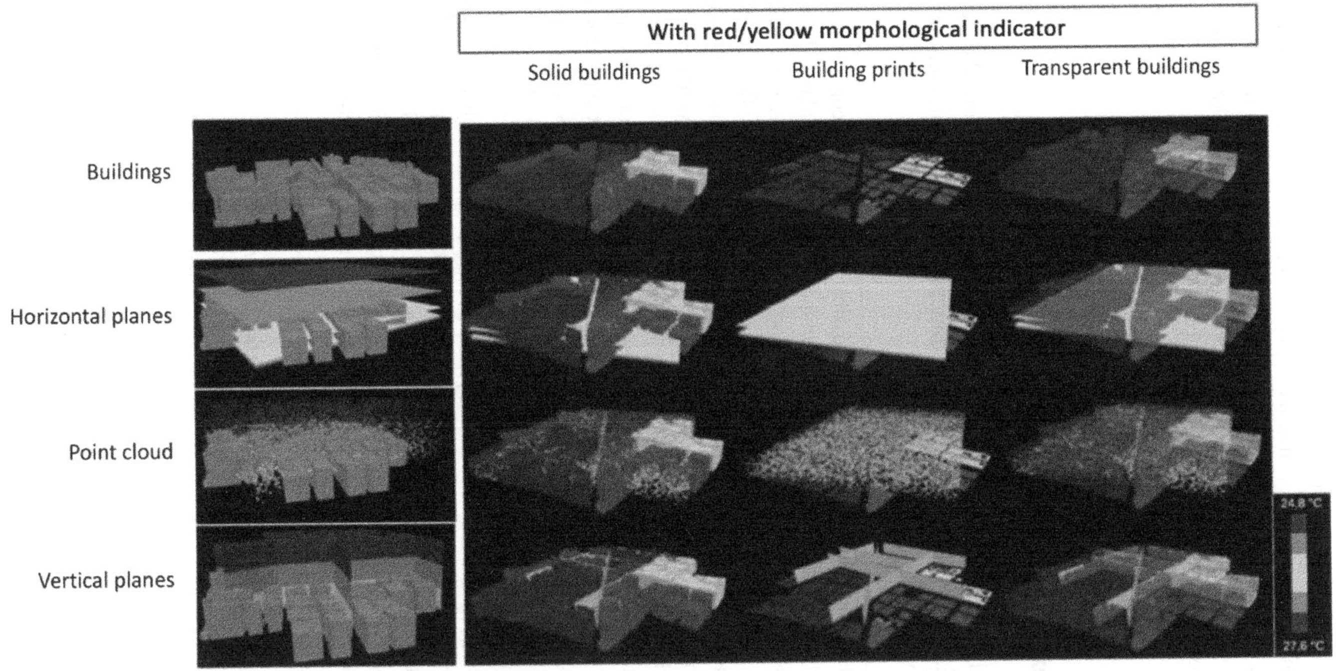

Figure 7.17. *Co-visualization of urban temperature and morphology data (Gautier et al. 2020). For a color version of this figure, see www.iste.co.uk/mericskay/cartographic.zip*

7.6. Conclusion

Geovisualization has always drawn on the knowledge, methods and practices of cartography, and continues to revisit existing conceptual and technical frameworks. Offering the possibility of interacting with data and their representation styles opens up a new field of possibilities for improving and, above all, rethinking (carto)graphic communication. This type of approach nevertheless requires a broad reflection from geovisualization designers, who must both think about and anticipate users and uses, while modeling functionalities and interfaces. This posture often implies taking a step back and moving toward a more interdisciplinary or even multidisciplinary manner with the targeted users. Putting the user back at the heart of geovisualization design proposals requires positioning oneself in view of the uses, as a designer, while mastering a set of knowledge and techniques from cartography, visual perception, graphic rendering, 2D and 3D spatial data manipulation, as well as computer development. Geovisualization designers hold the future of the tools they develop and provide to users in their hands, controlling the level of accessibility and simplification required. The objective in geovisualization is to offer users an optimal framework for cartographic communication, which is achieved by offering default proposals or functionalities that allow the user to explore, to analyze through interaction, or even to become a designer themselves. The user therefore depends to a large extent on the application designers, who choose to propose a certain number of representation modes and interaction functionalities to users.

Geovisualization and visual analysis aim to provide more assistance in the interpretation and explicability of spatiotemporal phenomena, as well as their models (Christophe 2020). Nevertheless, it would be foolish to believe that geovisualization tools, by their technical potential, are able to completely solve the problems of spatial analysis, (carto)graphic representation and decision-making on a spatiotemporal phenomenon. The "making sense" of data remains in the hands of the end user, who masters both the data and the represented phenomenon, especially when the user is the provider of the data and the analysis problems to be solved. It is necessary for designers to take care not to oversimplify, overinterpret, or even make decisions for the user in the most complex cases, but to give access to the points of view and the possibility of interpretation, as well as the associated uncertainty, with the help of keys to analysis and understanding. This is valid for experts, as well as for the general public, and sometimes requires additional explanations when opening an application, or when browsing the interfaces.

The abundance of data, uses and users leads us to rethink the challenges of information abstraction, simplification of (carto)graphic representations, and complexity of the spatiotemporal phenomena treated. Depending on the intended use and the user, from the same input data, it is possible to design (carto)graphic representations, whether they are simplified and synthetic representations to be potentially enriched with complementary information (the danger of a geophysical phenomenon, for the general public, for example), or whether they are multidimensional and expert representations (the visual analysis of scenarios of simulation of forecasts of a phenomenon, for experts, for example). This balance to be found and optimized remains at the heart of geovisualization research and puts forward many challenges to the extent that the general public is more and more accustomed to handling cartographic tools that are increasingly similar to those of experts.

In view of the complexity and richness of what geovisualization involves, it is essential to have a cross-cutting and integrative approach in order to be able to exploit all its components (data, style, interaction). In addition to "innovative" modes of cartographic representation, geovisualization has a pronounced tendency to offer users new ways of interacting with data that are ever closer to their digital uses, particularly on smartphones or tablets, and even immersion as a medium for an augmented reality user experience. The material advances of virtual reality helmets or augmented reality glasses already allow them to be used as information or communication mediums (virtual walks "in the past" or "in the future", virtual tours, museography, exploration of inaccessible locations, etc.). We are thus moving toward a flexible form of geovisualization tools, or at least toward an extension to new data sets, to new interaction tools, but also toward the possibility of moving from one type of representation to another, from one dimension to another, increasing the potential for information, communication and data analysis. This is why the design of dynamic interfaces, which integrates "situated" points of view on the data or the components of a phenomenon and leads us to look at a spatial, temporal or thematic dimension according to the user's actions, is a subject of research still to be developed. Having tools to manipulate all these dimensions opens up a large area for possible (carto)graphic representations for communication.

7.7. References

Andrienko, N. and Andrienko, G. (2006). *Exploratory Analysis of Spatial and Temporal Data: A Systematic Approach*. Springer, Berlin.

Beconyte, G. (2011). Cartographic styles: Criteria and parameters. In *Proceedings of the 25th International Cartographic Conference (ICA)*. Paris.

Bertin, J. (1967). *Sémiologie graphique ; les diagrammes, les réseaux, les cartes*. Mouton, Paris.

Blaise, J.Y., Davoine, P.A., Kaddouri, L., Mathian, H., Saint-Marc, C. (2015). Géovisualisation des temporalités des territoires : analyse comparative de la diversité des productions. In *Douzième rencontres Théo Quant*. Besançon.

Bonaccorsi, J., Jacquinod, F., Vogt, M. (2018). Regard pluridisciplinaire sur les usages sociaux de géovisualisations 3D pour la sensibilisation au risque d'inondation. *Revue internationale de géomatique*, 28(1), 69–94.

Brasebin, M., Christophe, S., Jacquinod, F., Vinesse, A., Mahon, H. (2016). 3D geovisualization & stylization to manage comprehensive and participative local urban plans. *ISPRS Annals of Photogrammetry, Remote Sensing & Spatial Information Sciences*, 4(2), 83–91.

Brewer, C.A. (1994). Color use guidelines for mapping and visualization. In *Visualization in Modern Cartography*, MacEachren, A.M. and Taylor, D.R.F. (eds). Elsevier Science, Amsterdam.

Card, S.K., Mackinlay, J.D., Shneiderman, B. (1999). *Readings in Information Visualization: Using Vision to Think*. Morgan Kaufmann, San Francisco.

Christophe, S. (2011). Creative Colours specification based on knowledge (COLorLE Gend System). *The Cartographic Journal*, 48, 138–145.

Christophe, S. (2012). Cartographic styles between traditional and original (towards a cartographic style model). In *AutoCarto Conference 2012*. Columbus.

Christophe, S. (2019). Color theory. In *The Geographic Information Science & Technology Body of Knowledge*, Wilson, J.P. (ed.) [Online]. Available at: https://gistbok.ucgis.org/bok-topics/color-theory [Accessed 8 July 2021].

Christophe, S. (2020). Geovisualization: Multidimensional exploration of the territory. *Proceedings of the 15th International Joint Conference on Computer Vision, Imaging and Computer Graphics Theory and Applications*, 3, 325–332.

Christophe, S. and Hoarau, C. (2012). Expressive map design based on Pop Art: Revisit of semiology of graphics? *Cartographic Perspectives*, 73, 61–74.

Christophe S., Duménieu B., Turbet J., Hoarau C., Mellado N., Ory J., Loi H., Masse A., Arbelot B., Vergne R. et al. (2016). Map style formalization: Rendering techniques extension for cartography. In *Expressive 2016: The Joint Symposium on Computational Aesthetics and Sketch-Based Interfaces and Modeling and Non-Photorealistic Animation and Rendering*, Bénard, P., Winnemöller, H. (eds). Eurographics Association, Goslar.

Davies, C., Fabrikant, S.I., Hegarty, M. (2015). Towards empirically verified cartographic displays. In *Cambridge Handbook of Applied Perception Research*, Szalma, J., Scerbo, M., Hancock, P., Parasuraman, R., Hoffman, R. (eds). Cambridge University Press, Cambridge.

Dey, A.K. (2001). Understanding and using context. *Personal and Ubiquitous Computing*, 5(1), 4–7.

Duménieu, B. and Christophe, S. (2016). Integrating expressive rendering techniques in a GIS cartographic pipeline. In *Agile 2016 Workshop "Code Loves Maps: Cartographically Oriented Programming Environments"*. Helsinki.

Dykes, J., MacEachren, A.M., Kraak, M.J. (2005). Advancing geovisualization. In *Exploring Geovisualization*, Dykes, J., MacEachren, A.M., Kraak, M.J. (eds). Pergamon Press, Amsterdam.

Ertz, O., Bocher, E., Sorensen, M., Sarafinof, D. (2020). OGC symbology conceptual model. Report, Open Geospatial Consortium.

Fabrikant, S. and Lobben, A. (2009). Introduction: Cognitive issues in geographic information visualization. *Cartographica*, 44, 139–143.

Fairbairn, D. (2006). Measuring map complexity. *The Cartographic Journal*, 43, 224–238.

Friendly, M. (2008). A brief history of data visualization. In *Handbook of Data Visualization*, Chen, C., Härdle, W.K., Unwin, A. (eds). Springer, Berlin.

Gautier, J., Davoine, P.A., Cunty, C. (2017). Visual search of cyclic spatio-temporal events. In *The 28th International Cartographic Conference ICC 2017*. Washington

Gautier, J., Brédif, M., Christophe, S. (2020). Co-visualization of air temperature and urban data for visual exploration. In *IEEE VIS Short Paper Proceedings*, IEEE Xplore, Salt Lake City, 71–75.

Geisthövel, R. and Hurni, L. (2018). Automated Swiss-style relief shading and rock hachuring. *The Cartographic Journal*, 55(4), 341–361.

Griffin, A.L. and Fabrikant, S.I. (2012). More maps, more users, more devices means more cartographic challenges. *The Cartographic Journal*, 49, 298–301.

Griffin, A.L., White, T., Fish, C., Tomio, B., Huang, H., Sluter, C.R., Bravo, J.V.M., Fabrikant, S.I., Bleisch, S., Yamada, M., Picanço, P. (2017). Designing across map use contexts: A research agenda. *International Journal of Cartography*, 3, 90–114.

Hägerstrand, T. (1970). What about people in regional. *Papers of the Regional Science Association Papers*, 24, 7–21.

Harrower, M.A. and Brewer, C.A. (2003). ColorBrewer.org: An online tool for selecting color schemes for maps. *The Cartographic Journal*, 40(1), 27–37.

Hoarau, C. and Christophe, S. (2017). Cartographic continuum rendering based on color and texture interpolation to enhance photo-realism perception. *ISPRS Journal of Photogrammetry and Remote Sensing*, 127, 27–38.

Jacquinod, F., Pedrinis, F., Edert, J., Gesquière, G. (2016). Automated production of interactive 3D temporal geovisualizations so as to enhance flood risk awareness. In *Proceedings of the Eurographics Workshop on Urban Data Modelling and Visualisation*. Liège.

Jégou, L. (2018). Expanding the Sémiologie Graphique for contemporary cartography, some ideas from visual semiotics, art history and design. *Cartography and Geographic Information Science*, 46(2), 182–188.

Jégou, L. and Deblonde, J.P. (2012). Vers une visualisation de la complexité de l'image cartographique. *Cybergeo: European Journal of Geography* [Online]. Available at: http://journals.openedition.org/cybergeo/25271 [Accessed 21 June 2021].

Jenny, B. and Hurni, L. (2006). Swiss-Style colour relief shading modulated by elevation and by exposure to illumination. *The Cartographic Journal*, 43(3), 198–207.

Kent, A. and Vujakovic, P. (2009). Stylistic diversity in European state 1:50 000 topographic maps. *The Cartographic Journal*, 46(3), 179–213.

Kerren, A., Stasko, J.T., Fekete, J.D., North, C. (2008). *Information Visualization: Human-Centered Issues and Perspectives*. Springer, Berlin.

Kraak, M.J. (2003). Geovisualization illustrated. *ISPRS Journal of Photogrammetry and Remote Sensing*, 57, 390–399.

Kraak, M.J. (2008). Geovisualization and time – New opportunities for the space-time cube. In *Geographic Visualization: Concepts, Tools and Applications*, Dodge, M., Turner, M., McDerby, M. (eds). Wiley, Chichester.

Lecomte, C., Vallée, J., Le Roux, G., Commenges, H. (2018). Le Mobiliscope, un outil de géovisualisation des rythmes quotidiens des métropoles. *Mappemonde*, 123 [Online]. Available at: https://journals.openedition.org/mappemonde/522.

Lloyd, D. and Dykes, J. (2011). Human-centered approaches in geovisualization design: Investigating multiple methods through a long-term case study. *IEEE Transactions on Visualization and Computer Graphics*, 17, 2498–2507.

Lobo, M.J., Appert, C., Pietriga, E. (2017). MapMosaic: Dynamic layer compositing for interactive geovisualization. *International Journal of Geographical Information Science*, 31(9), 1818–1845.

Loi, H., Hurtut, T., Vergne, R., Thollot, J. (2017). Programmable 2D arrangements for element texture design. *ACM Trans. Graph.*, 36(3), 1–17.

MacEachren, A.M. (1994). Visualization in modern cartography: Setting the agenda. In *Modern Cartography Series*, MacEachren, A.M. and Taylor, D.R.F. (eds). Pergamon Press, Oxford.

MacEachren, A.M. (1995). *How Maps Work: Representation, Visualization, and Design*. Guilford Publications, New York.

MacEachren, A.M. and Kraak, M.J. (2001). Research challenges in geovisualization. *Cartography and Geographic Information Science*, 28, 3–12.

Masse, A. and Christophe, S. (2016). Améliorer la perception du réalisme dans la géovisualisation du littoral ; utilisation de données spatio-temporelles hétérogènes. *Revue internationale de géomatique*, 26(4), 403–424.

Mericskay, B. (2016). La cartographie à l'heure du géoweb ; retour sur les nouveaux modes de représentation spatiale des données numériques. *Cartes et géomatique*, 229/230, 37–50.

Mericskay, B. (2021). La géovisualisation de données massives sur le Web ; entre avancées technologiques et évolutions cartographiques. *Mappemonde*, 131 [Online]. Available at: https://journals.openedition.org/mappemonde/5595.

Mellado, N., Vanderhaeghe, D., Hoarau, C., Christophe, S., Brédif, M., Barthe, L. (2017). Constrained palette-space exploration. *ACM Transactions on Graphics*, 36(4), 1–14.

Morrison, J.L. (1974). A theoretical framework for cartographic generalization with the emphasis on the process of symbolization. *International Yearbook of Cartography*, 14, 115–127.

Nöllenburg, M. (2007). Geographic visualization. In *Human-Centered Visualization Environments*, Kerren, A., Ebert, A., Meyer, J. (eds). Springer, Berlin.

Ooms, K., Cöltekin, A., De Maeyer, P., Dupont, L., Fabrikant, S.I., Incoul, A., Kuhn, M., Slabbinck, H., Vansteenkiste, P., Van der Haegen, L. (2015). Combining user logging with eye tracking for interactive and dynamic applications. *Behavior Research Methods*, 47, 977–993.

Ory, J. (2017). Cartes à la une : les cartes topographiques ont du style ! *GéoConfluences* [Online]. Available at: http://geoconfluences.ens-lyon.fr/informations-scientifiques/a-la-une/carte-a-la-une/les-carte-topo-ont-du-style [Accessed 21 June 2021].

Ory, J., Christophe, S., Fabrikant, S.I., Bucher, B. (2015). How do map readers recognize a topographic mapping style? *The Cartographic Journal*, 52(2), 193–203.

Ory, J., Christophe, S., Fabrikant, S.I., Bucher, B. (2018). Modèle de conception cartographique intégrant l'expérience utilisateur. In *CIST 2018, Représenter les territoires*. Rouen.

Palsky, G. (2021). Carte. *HyperGéo* [Online]. Available at: http://www.hypergeo.eu/spip.php?article266 [Accessed 21 June 2021].

Perrin, O., Christophe, S., Jacquinod, F., Payrastre, O. (2020). Visual analysis of inconsistencies in hydraulic simulation data. In *XXIV International Society for Photogrammetry and Remote Sensing Congress*. Nice.

Pietriga, E., Bau, O., Appert, C. (2010). Representation-independent in-place magnification with sigma lenses. *IEEE TVCG*, 16(3), 455–467.

Pindat, C., Pietriga, E., Chapuis, O., Puech, C. (2012). JellyLens: Content-aware adaptive lenses. In *Proceedings of the 25th Symposium on User Interface Software and Technology*. Cambridge, 261–270.

Poorthuis, A., Van der Zee, L., Guo, G., Keong, J.H., Dy, B. (2020). Florence: A web-based grammar of graphics for making maps and learning cartography. *Cartographic Perspectives*, 96, 32–50.

Post, F., Nielson, G., Bonneau, G.P. (2003). *Data Visualization: The State of the Art*. Springer, New York.

Ritschard, L., Georis-Creuseveau, J., Gourmelon, F. (2016). Information géographique numérique et gestion des territoires. Application aux zones côtières françaises. In *Conférence ; intelligences numériques*. Quebec.

Roth, R. (2012). Cartographic interaction primitives: Framework and synthesis. *The Cartographic Journal*, 49(4), 376–395.

Roth, R., Cöltekin, A., Delazari, L., Fonseca, F.H., Griffin, A., Hall, A., Korpi, J., Lokka, I., Mendonça, A., Ooms, K., Elzakker, C. (2017). User studies in cartography: Opportunities for empirical research on interactive maps and visualizations. *International Journal of Cartography*, 3, 1–29.

Saint-Julien, T. (2001). Diffusion spatiale et modélisation du changement. In *Modèles en analyse spatiale*, Sanders, L. (ed.). Hermès, Paris.

Slocum, T.A., Blok, C., Jiang, B., Koussoulakou, A., Montello, D.R., Fuhrmann, S., Hedley, N.R. (2001). Cognitive and usability issues in geovisualization. *Cartography and Geographic Information Science*, 28, 61–75.

Touya, G., Christophe, S., Hoarau, C. (2016). Clutter and map legibility in automated cartography: A research agenda. *Cartographica*, 51(4), 198–207.

Tversky, B. (2005). Visuospatial reasoning. In *The Cambridge Handbook of Thinking and Reasoning*, Holyoak, K. and Morrison, R. (eds). Cambridge University Press, Cambridge.

Wilkinson, L. (2005). *The Grammar of Graphics*. Springer, Berlin.

Willats, J. and Durand, F. (2005). Defining pictorial style: Lessons from linguistics and computer graphics. *Axiomathes*, 15(3), 319–351.

Conclusion

Boris MERICSKAY
ESO, University of Rennes 2, France

The numerous technical advances in the representation of spatial data and the new uses of maps are at the center of important challenges in the transmission of information. The seven chapters of this book have, in these terms, largely put forward the need to question and reconsider cartographic communication with regard to the major (r)evolutions (technological, methodological, informational) that the map, as an object of mediation, has undergone since the early 2000s.

In this era of the Geoweb, collaborative databases or social networks, the unidirectional communication model where the cartographer (as the sender) produces a "finalized" map for an audience (as the receiver) is no longer as clear, and deserves to be reconsidered. As amply presented and exemplified in the various contributions of the book, a multitude of cartographic communication models coexist today, mobilizing heterogeneous approaches, tools and audiences, in ever more diversified contexts of use. The construction of a map, from its design to its dissemination, is no longer linear and predefined, insofar as all the steps are intertwined and there is a confusion around the parties involved.

However, even though the dynamics and contemporary modalities of information transmission through maps tend to be more complex because they are numerous and protean, several major trends exist and deserve to be exposed and formalized as a basis for future research.

First, the question of technical instrumentation, which tends to garner more and more importance in training courses as well as in practices, is only of secondary

Cartographic Communication,
coordinated by Boris MERICSKAY.

importance. The digitalization of cartography, especially on the web, does not really constitute a communication issue. Regardless of the fact that the plethora of tools, as well as the multitude of data, authorize ever more original and varied modes of representation, cartography remains above all a question of choice on the designer's part. In some cases, the use of complex tools will be able to meet certain specific needs, but in other contexts, the use of very simple tools will be more than sufficient, as they are adapted to the objectives targeted by the map.

Many software publishers and mapping service providers also tend to suggest that in order to represent the volume and complexity of spatial data that proliferate today, "modern" instrumentation is accompanied by new logic and new dedicated (carto)graphic languages. However, in practice, the sign systems conceptualized and formalized more than 50 years ago, such as graphic semiology, do not disappear in favor of new approaches and methods. On the contrary, it is clear that the knowledge and understanding of these fundamentals are more relevant than ever. These visual grammars are indeed widely used by cartographers, data journalists and web designers, who continue to use them and make them evolve within new technical environments, in multiple contexts of use.

In the line of thought around the encoding of information by sign systems, the question of map "efficiency" is becoming increasingly significant, especially in a context where anyone can access a multitude of geographic data, produce diverse and varied maps, and disseminate them to the greatest number of people. In this "competitive" environment around spatial representations, cartographic communication requires both a certain respect for cartographic conventions and the use of new modes of representation (multi-representation, webmapping, geovisualization), which must be adapted as much as possible to the uses and profiles of the targeted audiences.

Rethinking the role of maps, as well as the status of the public, is at the core of major challenges today, in order to apprehend and understand the mechanisms of information mediation through maps. The receiver of a map is no longer just a user who passively receives a representation, but instead an active user who can handle these productions, modify them, share them and even build his own maps via dedicated applications.

The growing interactivity between data, maps and the user requires cartographic communication to be reconsidered, especially in relation to the development of geovisualization. The static map, which is still the main way to transmit a message, is gradually giving way to dynamic and interactive applications, where the objective is to allow the user to explore data sets and to design spatial representations in line with his/her practices.

This transformation of the role and status of map users is now accentuated by the development of map personalization. Whether it is a question of displaying data according to his habits or location, or adapting the symbology of the data or the style of the base map according to his profile or uses, the user plays a central role in cartographic communication at all stages of map production. This tendency is also largely questioned by new devices for visualization and map use, such as smartphone screens, augmented reality glasses or virtual reality headsets, which are changing the modalities of transmission of geographical information.

Lastly, from the perspective of understanding the processes and challenges of cartographic communication, the deepening of knowledge through the prism of disciplines such as information and communication sciences, computer graphics or data visualization allows for reflection from a theoretical, technical and methodological point of view. The seven chapters of this book have thus shown an interest in using other approaches and crossing viewpoints in order to understand the principles and the challenges of cartographic communication in the 21st century in a more detailed manner.

List of Authors

Françoise BAHOKEN
AME-SPLOTT
IFSTTAR
Géographie-cités
Gustave-Eiffel University
Marne-la-Vallée
France

Colette CAUVIN-REYMOND
LIVE, CNRS
University of Strasbourg
France

Sidonie CHRISTOPHE
LASTIG
ING
ENSG
Gustave-Eiffel University
Saint-Mandé
France

Timothée GIRAUD
RIATE
CNRS
University of Paris-Diderot
France

Laurent JÉGOU
LISST
University of Toulouse-Jean Jaurès
France

Nicolas LAMBERT
RIATE, CNRS
University of Paris-Diderot
France

Boris MERICSKAY
ESO
University of Rennes 2
France

Matthieu NOUCHER
Passages, CNRS
University of Bordeaux
France

Christine ZANIN
Géographie-cités
University of Paris-Diderot
France

Ronan YSEBAERT
RIATE, CNRS
University of Paris-Diderot
France

Index

D, E, F

G, I

M, O, P

ted and bound by CPI Group (UK) Ltd, Croydon, CR0 4YY

16/09/2026

14963201-0003